Arsenic in the Environment - Sources, Impacts and Remedies

Edited by S.M. Imamul Huq

Published in London, United Kingdom

Arsenic in the Environment - Sources, Impacts and Remedies
http://dx.doi.org/10.5772/intechopen.1001534
Edited by S.M. Imamul Huq

Contributors
Amartya De, Nilanjana Roy, Jiang Wu, Zhenzhen Guan, Yang Ling, Sweety Nath Barbhuiya, Ringthardi Warisa, Dharmeswar Barhoi, Jurimoni Das, Sarbani Giri, Hiroshi Yamauchi, Takahiko Yoshida, Ayako Takata, Ayesha Malik, Fakhira Khalid, Nigah Hidait, Khalid Mehmood Anjum, Saima Mahad, Abdul Razaq, Hamda Azmat, Muhammad Bilal Bin Majeed, Mangalathettu Binumon Thankachan, Gayathri S. Kamath, Greeshma Sasikumar, Sreejith Parameswara Panicker, Francisco Jose Alguacil, Jose Ignacio Robla

First published in London, United Kingdom, 2023 by IntechOpen
IntechOpen is the global imprint of INTECHOPEN LIMITED, registered in England and Wales, registration number: 11086078, 5 Princes Gate Court, London, SW7 2QJ, United Kingdom

British Library Cataloguing-in-Publication Data
A catalogue record for this book is available from the British Library

Additional hard and PDF copies can be obtained from orders@intechopen.com

Arsenic in the Environment - Sources, Impacts and Remedies
Edited by S.M. Imamul Huq
p. cm.
Print ISBN 978-1-83769-655-0
Online ISBN 978-1-83769-654-3
eBook (PDF) ISBN 978-1-83769-656-7

Meet the editor

A renowned educationist and world-reputed scientist, Prof. Dr. S.M. Imamul Huq is a pioneer in research on arsenic in the environment and its transfer to the food chain He is trilingual and a part-time French language teacher at the Institute of Modern Languages, University of Dhaka, Bangladesh. He has more than 300 publications to his credit and was awarded the Bangladesh Academy of Sciences Gold Medal for his outstanding contribution to the agricultural and biological sciences. He is also the recipient of the Bangladesh UGC Award in the Physical and Biological Sciences/Agricultural Sciences. He was elected as a "Distinguished Alumni" by the Asian Institute of Technology Alumni Association (AITAA) in 2008, and the best educationist by the Ministry of Agriculture and the Food and Agriculture Organization of the United Nations on World Soil Day 2022.

Contents

Preface

Environmental contamination has become a common phenomenon in the developing world because of growing urbanization and industrialization, which in many instances pollute the environment. Such pollution, either through anthropogenic activities or natural contaminants, has led to great adverse effects for humans. Besides organic pollutants, some heavy metals and metalloids are also toxic. Arsenic is one such metalloid, which is ubiquitous in nature and has gained prominence due to its toxic effects on humans, animals, and plants. This book discusses the impacts of this carcinogen and the progress made thus far in controlling and remediating contamination/pollution by arsenic. The book contains seven chapters contributed by authors from around the globe. The chapters are organized into three sections: "Sources" "Impacts" and "Remedies".

Section 1 includes two chapters. Chapter 1 discusses the interaction of elements in soil, water, and air with arsenic. It describes the remedial actions of two native plants on arsenic-infested rats and the authors suggest the same approach for future study in human beings. Chapter 2 discusses the emission of arsenic in the flue gas in coal mines and coal-based industries and the various possibilities of abating/controlling the release of the toxin into the environment.

Section 2 includes four chapters. Chapter 3 reviews the effects of arsenic in soil on plant functions and plant health. Chapter 4 examines the effect of inorganic arsenic and its methylation on human health. It discusses the conversion of inorganic arsenic to various organic arsenic compounds in the human body and how these compounds affect human health. The chapter also presents comparative analytical perspectives on methylated arsenic products. It also discusses the use of arsenic in chemical warfare. Furthermore, the chapter evaluates the adverse effects of arsenic poisoning, such as carcinogenesis and damage to the brain and nervous system. The authors raise an important question about the consumption of rice, as it has been reported that rice is an arsenic accumulator. Chapter 5 describes various aspects of arsenic poisoning on fish, detailing its physical, biological, biochemical, hematological, and other effects. Chapter 6 reviews the effect of arsenic exposure and its carcinogenic effects on the skin. The chapter examines biochemical reactions, genetic relations, sexual differences, and more.

Section 3 includes one chapter. Chapter 7 addresses the remediation possibilities for arsenic contamination and its consequences. It discusses the removal of arsenic by biosorbent methods and compares these methods' effectiveness.

I would like to thank the staff at IntechOpen, especially Ivana Barac, for their invitation to edit this book and their support throughout the publication process.

S.M. Imamul Huq
Department of Soil, Water and Environment,
University of Dhaka,
Dhaka, Bangladesh

Section 1

Sources

Chapter 1

Consequences of Arsenic in the Environment

Amartya De and Nilanjana Roy

Abstract

Throughout the environment, including in food, water, soil, and the air, arsenic is a naturally occurring metalloid. An important global problem is the human exposure to arsenic from numerous sources, such as polluted groundwater and other human activities. It poses a serious danger to people's health, economic security, and social standing, particularly in the world's least developed nations, as its exposure to humans has been related to a wide range of illnesses. Diabetes, hyperkeratosis, cancer, hypertension, and neurodegeneration are a few of the serious illnesses that have been connected to arsenic exposure. There is currently no known or authorized treatment for arsenic poisoning. We made an effort to shed light on some of the most important scientific facts on arsenic toxicity that have been published in the literature. It is important for policymakers to develop regulations for a cleaner environment and raise public knowledge of arsenic toxicity.

Keywords: arsenic, *Bauhinia acuminata*, *Glycosmis pentaphylla*, toxicity, rat

1. Introduction

Arsenic is a toxic element that occurs naturally in the environment and can also be released through human activities such as mining, smelting, and agriculture. This element can have serious health implications, including cancer, skin lesions, and cardiovascular disease. Arsenic contamination in drinking water and food sources is a major concern, especially in developing countries.

This paper aims to provide an overview of sources, implications, and remedies related to arsenic in the environment. The sources of arsenic are discussed, including natural sources, such as geological formations, and human activities, such as mining and pesticide use. The implications of arsenic exposure are also examined, including its effects on human health, agriculture, and ecosystems.

Several remediation techniques are discussed, including physical, chemical, and biological methods. These techniques include the use of adsorbents, coagulants, ion exchange, and biological treatment. Additionally, policy measures and regulations related to arsenic in the environment are reviewed.

Overall, this paper highlights the importance of understanding the sources, implications, and remediation of arsenic in the environment. Effective measures are needed to mitigate the risks associated with arsenic exposure and protect public health and the environment.

2. Chemistry of origin

Arsenic is a naturally occurring element that is present in soil, water, and air. It is widely distributed in the earth's crust and is found in various minerals such as arsenopyrite, realgar, and orpiment. Arsenic is a toxic substance that can have serious health effects on humans and animals, including cancer, skin lesions, and developmental problems. In this response, we will explore the chemistry of the origin of arsenic in the environment [1].

There are several sources of arsenic in the environment, including natural and anthropogenic sources. Natural sources of arsenic include volcanic emissions, weathering of rocks, and the breakdown of organic matter. Arsenic can also be released into the environment through anthropogenic activities such as mining, smelting, and the use of arsenic-containing pesticides [2].

One of the primary factors that determine the mobility of arsenic in the environment is its oxidation state. Arsenic exists in two primary oxidation states, arsenite (As(III)) and arsenate (As(V)). Arsenite is more toxic and mobile than arsenate, and is more commonly found in reducing environments such as groundwater and sediments. Arsenate is more stable and less toxic than arsenite, and is more commonly found in oxidizing environments such as surface water and soils [3].

The mobility and toxicity of arsenic are also influenced by the pH of the environment. In acidic environments, arsenic tends to be more mobile and bioavailable, while in alkaline environments, arsenic is less mobile and less toxic. The presence of certain minerals, such as iron and manganese oxides, can also influence the mobility and bioavailability of arsenic in the environment [4].

Overall, the chemistry of the origin of arsenic in the environment is complex and influenced by a variety of factors. Understanding the sources, mobility, and toxicity of arsenic is critical for developing effective strategies to mitigate its impact on human and environmental health.

3. Accumulation in plants

Plants are important vectors for arsenic accumulation in the environment, as they can take up arsenic from the soil and water and translocate it to their tissues. Arsenic uptake by plants occurs through the roots, and the amount of arsenic that a plant can accumulate depends on a variety of factors, including the concentration of arsenic in the soil and water, the pH of the soil, and the plant species [5].

Some plant species have developed mechanisms to tolerate and accumulate high levels of arsenic in their tissues, a process known as arsenic hyperaccumulation. These plants can be used for phytoremediation of arsenic-contaminated soils and water, as they can effectively remove arsenic from the environment [6].

Several studies have investigated the mechanisms of arsenic accumulation in plants and have identified a range of genes and transporters involved in arsenic uptake, translocation, and detoxification. For example, the arsenate reductase gene (ACR2) has been shown to play a key role in reducing arsenate to arsenite, which is more easily taken up by plant roots. The phosphate transporter gene (PHO1) has also been implicated in arsenic uptake, as it can transport both phosphate and arsenate into plant cells [7].

In addition to phytoremediation, arsenic hyper accumulating plants also have potential applications in agriculture, as they can be used to improve the arsenic tolerance of crop plants. For example, transgenic rice plants that overexpress the arsenic tolerance gene (OsHAC1;1) have been shown to have increased arsenic tolerance and reduced arsenic accumulation in their tissues [8].

4. Plant health impact

In plants, arsenic can cause a wide range of adverse effects, including chlorosis, stunted growth, and reduced photosynthesis. Arsenic can also disrupt the plant's water uptake and nutrient absorption, leading to nutrient deficiencies and plant death. Additionally, arsenic can accumulate in the plant tissue, making it toxic to herbivores that feed on the plant [9].

One of the primary ways that plants are exposed to arsenic is through contaminated soil and water. In areas where arsenic-containing pesticides were heavily used, soil and water can become contaminated, and plants growing in that area can take up the toxic metalloid through their roots. Arsenic can also be absorbed by plants through their leaves if it is present in the air as a result of industrial emissions or volcanic activity [10].

The impact of arsenic on plant health can have significant economic consequences, particularly in agricultural areas. Arsenic contamination in crops can reduce crop yields and quality, affecting the income of farmers and the availability of food for consumers. Additionally, the consumption of arsenic-contaminated food can have severe health consequences for humans and animals [11].

To mitigate the impact of arsenic on plant health, various strategies have been proposed, including phytoremediation, which involves using plants to remove arsenic from contaminated soil and water. Other approaches include the use of microorganisms to degrade or immobilize arsenic in soil, and the development of arsenic-resistant crops through genetic engineering [12].

5. Toxic levels

Exposure to arsenic can occur through inhalation, ingestion, or absorption through the skin. Arsenic toxicity can cause a range of health effects, including skin lesions, cancer of the skin, lung, bladder, and kidney, cardiovascular disease, and neurological effects.

In many parts of the world, arsenic contamination of drinking water is a major public health concern. The World Health Organization (WHO) has set a guideline value of 10 μg of arsenic per liter of drinking water to protect against the health effects of long-term exposure [13].

Several countries, including Bangladesh, India, and parts of Latin America, have been severely affected by arsenic contamination of drinking water, with millions of people exposed to toxic levels of arsenic. In Bangladesh alone, it is estimated that up to 77 million people are at risk of arsenic contamination of drinking water [14].

Efforts to reduce arsenic exposure have focused on improving access to safe drinking water through the development of arsenic removal technologies, such as reverse osmosis, and the promotion of safe water sources [15].

In addition to drinking water, arsenic exposure can also occur through the consumption of contaminated food, particularly rice, which can absorb arsenic from soil and water. The WHO has set a maximum limit of 0.2 mg of arsenic per kilogram of rice to protect against the health effects of long-term exposure [16].

6. Biosorption

Biosorption is a process in which living or non-living biomass is used to remove contaminants from the environment. In recent years, biosorption has emerged as an effective and environmentally friendly technology for the removal of arsenic from contaminated water.

Biosorption of arsenic involves the use of biological materials, such as microorganisms, algae, and plants, to remove arsenic from contaminated water. These materials have the ability to bind arsenic to their cell walls or to other cellular components, effectively removing it from the water.

Several studies have demonstrated the effectiveness of biosorption for the removal of arsenic from contaminated water. For example, a study conducted in Bangladesh showed that the use of the aquatic plant water hyacinth effectively removed arsenic from contaminated water, reducing arsenic levels by up to 90% [17]. Another study conducted in India showed that the use of the fungus *Aspergillus niger* was effective in removing arsenic from contaminated water, reducing arsenic levels by up to 70% [18].

Biosorption has several advantages over other methods of arsenic removal, such as chemical precipitation and membrane filtration. Biosorption is a low-cost and environmentally friendly technology that does not require the use of chemicals or energy-intensive processes. Biosorption is also effective over a wide range of pH and temperature conditions, making it suitable for use in a variety of environments.

7. Soil biodiversity

Soil biodiversity plays a crucial role in maintaining ecosystem functioning, and its role in mitigating heavy metal pollution has received significant attention in recent years. Arsenic (As) is a toxic metalloid that is widely distributed in the environment due to both natural and anthropogenic sources. Arsenic pollution in soil is a major concern worldwide, particularly in developing countries where industrialization and agricultural practices have led to significant contamination of soils with As.

Soil biodiversity plays a crucial role in mitigating As pollution in soil. Studies have shown that certain microbial communities in soil, such as fungi and bacteria, are capable of transforming As into less toxic forms or immobilizing As in soil, thereby reducing its mobility and bioavailability to plants and other organisms [19–21].

Moreover, soil biodiversity can also play a crucial role in phytoremediation of As-contaminated soils. Certain plant species are capable of accumulating high levels of As in their tissues without showing any toxic effects. Soil biodiversity, particularly the microbial communities associated with plant roots, can enhance the uptake and accumulation of As in plant tissues, thereby facilitating phytoremediation of As-contaminated soils [22, 23].

8. Ecosystem dynamics

Ecosystem dynamics of arsenic in the environment refers to the complex interactions between different components of an ecosystem, including living organisms, water, soil, and air, that influence the fate, transport, and bioavailability of arsenic.

The uptake and accumulation of arsenic by plants and animals is also influenced by various factors, including the type of organism, the concentration and form of arsenic in the environment, and the presence of other nutrients and contaminants. Some organisms, such as certain types of algae and bacteria, are able to metabolize or detoxify arsenic, while others, such as humans and animals, are susceptible to its toxic effects [24].

Understanding the ecosystem dynamics of arsenic is important for developing strategies to mitigate its impacts on human health and the environment. This can include approaches such as remediation of contaminated soils and water, monitoring of exposure levels, and development of crop varieties that are less susceptible to arsenic uptake [25].

9. Sequestration in crops

Sequestration refers to the process of isolating or removing a substance from its environment and placing it in a different location where it is less harmful. In the case of crops, sequestration of arsenic is an important mechanism for reducing the concentration of this toxic element in the environment.

Crops play an important role in sequestering arsenic from the environment. This is because plants have the ability to absorb and accumulate metals from the soil, water, and air. Once absorbed, these metals can be sequestered in the plant tissues, including the roots, stems, leaves, and fruits, and removed from the environment. The process of arsenic sequestration in crops is complex and depends on several factors, including the type of crop, the soil conditions, and the concentration of arsenic in the environment [26].

Several studies have investigated the ability of different crops to sequester arsenic from the environment. For example, research has shown that rice is a highly efficient crop for sequestering arsenic from contaminated soils and water. Rice plants have the ability to absorb arsenic through their roots and transport it to the shoots, where it can be sequestered in the grain. However, the accumulation of arsenic in rice grain is a major concern for food safety, as it can pose a risk to human health [27].

Other crops that have been shown to be effective at sequestering arsenic from the environment include sunflowers, Indian mustard, and ferns. These crops have the ability to hyperaccumulate arsenic, meaning that they can accumulate much higher concentrations of arsenic than other plants. This makes them useful for phytoremediation, which is the use of plants to remove contaminants from the environment [28].

10. Abatement of toxicity

One of the most effective strategies to abate arsenic toxicity is to remove it from contaminated sites. Various techniques have been developed for arsenic removal, such as coagulation/flocculation, adsorption, ion exchange, membrane filtration, and biological methods [29]. These methods are based on the physical and chemical properties of arsenic and can be applied to different types of contaminated media, such as water, soil, and sediment.

Another strategy for abating arsenic toxicity is to prevent exposure to it. This can be achieved by implementing strict regulations on industrial processes that release arsenic into the environment, promoting safe disposal of arsenic-containing waste, and providing safe drinking water to communities that are at risk of arsenic exposure [29–33].

In addition to these strategies, bioremediation has also been explored as a potential method for abating arsenic toxicity. Bioremediation involves the use of microorganisms to transform or remove toxic pollutants from the environment [34]. For example, some bacteria and fungi are capable of reducing arsenic to less toxic forms, such as arsenite (As III) or arsine gas (AsH3), through a process called biotransformation [35]. Other microorganisms can oxidize arsenic to less soluble and less mobile forms, such as arsenate (As V), which can then be immobilized in the soil [36].

11. Preventive measures

Here are some preventive measures to mitigate arsenic contamination in the environment:

Monitoring and testing: Regular monitoring and testing of water sources, soil, and air quality can help detect and prevent arsenic contamination. Testing can be conducted using various techniques, including colorimetric methods, atomic absorption spectrometry, and inductively coupled plasma mass spectrometry.

Environmental regulations: Strict environmental regulations and guidelines can help prevent arsenic contamination in the environment. These regulations can include restrictions on the use of arsenic-containing pesticides, the discharge of industrial effluents, and the disposal of arsenic-containing waste.

Alternative agricultural practices: Alternative agricultural practices such as crop rotation, phytoremediation, and organic farming can reduce the use of arsenic-containing pesticides and fertilizers, which can lead to reduced arsenic contamination in soil and water.

Remediation techniques: Various remediation techniques such as electrocoagulation, adsorption, and membrane filtration can be used to remove arsenic from contaminated water sources.

Public awareness: Raising public awareness about the health risks associated with arsenic exposure and the importance of preventing contamination can help reduce arsenic exposure in communities [37–39].

12. Response to arsenic

Efforts to address the problem of arsenic contamination in the environment are ongoing. Strategies include the development of new technologies for removing arsenic from water, the implementation of better agricultural practices to reduce arsenic contamination in soil, and the regulation of industrial processes to minimize arsenic emissions. However, much more work is needed to address this serious environmental and public health problem [40].

The results on study with rats conducted by the authors are presented below.

13. Process of developing arsenic toxicity in rats

The process of developing arsenic toxicity in rats typically involves exposing them to arsenic through various routes such as oral ingestion, inhalation, or injection. The duration and dose of exposure can vary depending on the study design.

Ingestion of arsenic in drinking water is a common route of exposure in many studies. Rats may be given water containing different concentrations of arsenic, and the effects on their health and organ function are observed over time.

The toxicity of arsenic can affect various organs and systems in rats, including the liver, kidneys, heart, lungs, and immune system. The toxic effects of arsenic can manifest as inflammation, cell damage, and oxidative stress, leading to tissue damage and dysfunction.

To determine the development of arsenic toxicity, various techniques may be used to measure the levels of arsenic in the blood, urine, or tissues of rats. Additionally, histological examination of the organs can reveal the presence of arsenic-induced damage.

Overall, the process of developing arsenic toxicity in rats involves exposing them to arsenic through various routes, monitoring their health and organ function over time, and assessing the toxic effects of arsenic through various techniques.

14. Methodology

14.1 Chemicals used

The following companies: Bangalore Genei (India), Cogent (India), Merk (Germany), Promega (USA), Rankem (India), Biovision (USA), and Sigma Chemicals provided all of the chemicals and kits utilized in this investigation (USA).

Plant identification was done from BSI, the plants were recognized (Botanical Survey of India, Howrah, Kolkata). For *Bauhinia acuminata*, the specimen voucher number was WBUAFS/Kol/2, while for *Glycosmis pentaphylla*, it was WBUAFS/Kol/1.

14.2 Extract preparation

Preparation of *Bauhinia acuminata* stembark extract with *Glycosmis pentaphylla* leaf extract.

Fresh *Glycosmis pentaphylla* plant leaves and *Bauhinia acuminata* stem barks were gathered from the neighborhood. Afterwards, the leaves and barks were divided into pieces, cleaned with distilled water, and allowed to dry for 7 days in a shaded area. With a grinding device, the dried leaves and barks were ground into a coarse powder. Using a soxhlet device, the powder was extracted using methanol (99%) for *G. pentaphylla* and ethanol (99%) for *Bauhina accuminata* while the procedure was run for 8 to 12 h. Following extraction, a rotary evaporator was used to condense the whole solution. For 2 to 3 days, the condensed solution was left at room temperature. For later usage, they were dissolved in water (triple-distilled water).

A stock solution of 100 ppm of carbazole (purity 98.7%) procured from M/S Sigma Aldrich was prepared in methanol as standard. The HPLC Agilent Technologies 1200 series, coupled with PDA detector was used for carbazole estimation.

A stock solution of 100 ppm of Baicalein(purity 98) procured from M/S Sigma Aldrich was prepared in ethanol as standard. The HPLC Agilent Technologies 1200 series, coupled with PDA detector was used for Baicalein estimation.

14.3 Powder preparation

Preparation of *Bauhinia acuminata* stem bark powder and *Glycosmis pentaphylla* leaf powder.

Fresh *G. pentaphylla* plant leaves and *B. acuminata* stem bark were obtained from the neighborhood, cleaned with water, cut into pieces, and dried in the shade for 7 days.

The dried leaves and barks were ground into a coarse powder in a mill, dissolved in warm water of the Millipore quality, and then given to the test animals after filtering.

14.4 Animal experimentation

72 mature albino rats of either sex, weighing 150–200 g, were purchased from a licensed animal breeder. They spent the 7 days leading up to the trial being acclimated in the experimental animal room while living in polypropylene cages. Standard pellet food and unlimited water were given to the animals. The technical program, videno EC/235/2013/CPCSEA, dated September 5, 2013, was authorized by the institution's animal ethical committee.

14.5 Laboratory design

There were two sections to the entire experiment.

14.5.1 Experiment A

Six groups, Go, G1, G2, G3, G4, and G5, each with 12 rats, were formed from 72 animals of either sex, weighing 150–200 g. Go group rats were given unlimited access to food and water. Sodium arsenite, dissolved in distilled water, was given to rats in groups G1, G2, G3, G4, and G5 at a dose of 4 mg/kg daily for 90 days. Animals in groups G2 and G3 received doses of 320 and 160 mg/kg body weight, or 0.544 and 0.272 mg of carbazole per kilogram, respectively, from days 91 to 120. Animals in groups G4 and G5 received ethanolic extract of *B. acuminata* diluted in distilled water at doses of 300 and 150 mg/kg b.w. (equal to 0.99 and 0.495 mg of baicalein per kg, respectively) from day 91 to day 120. This set of animals served as the experimental control group since they were not given the plant extract.

For 120 days, Go(12) Control got water.

G1, (12) was given sodium arsenite for 90 days and water for 30 days.

G2(12) got sodium arsenite for 90 days and GP 320 mg/kg for 30 days.

G3, (12) was given sodium arsenite for 30 days and GP 160 mg/kg for 90 days.

G4. (12) was given sodium arsenite for 90 days and BA 300 mg/kg for 30 days.

G5:(12) got sodium arsenite for 90 days and BA 150 mg/kg for 30 days.

14.5.2 Experiment B

Six groups, Ao, A1, A2, A3, A4 and A5, each containing 12 rats, were formed from 72 animals at random. Rats in group A were given unlimited access to food and water. For 90 days, rats in groups A1, A2, A3, A4 and A5received sodium arsenite at a dose of 4 mg/kg per day in drinking water. From days 91 to 120, animals in groups A2 and A3 received leaf powder of *G. pentaphylla* dissolved in distilled water at doses of 500 and 250 mg/kg (1/10th and 1/20th of LD50, respectively). From day 91 to day 120, animals in groups A4 and A5 received stem bark powder from *B. acuminata* dissolved in distilled water at doses of 350 and 175 mg/kg (1/10"" and 1/20"" of LD50), respectively.

Each group of animals served as the experimental control by not receiving the powder treatment.

Water was provided to Ao(12) Control for 120 days.

A1, (12) was given sodium arsenite for 90 days and water for 30 days.

A2, (12) got arsenite for 90 days and GP powder 500 mg/kg for 30 days.
For 90 days, A3, (12) got sodium arsenite, and 250 mg/kg of GP powder for 30 days.
A4(12) got sodium arsenite for 90 days and BA powder 350 mg/kg for 30 days.
A5(12), During 90 days, got sodium arsenite, and 175 mg/kg of BA powder for 30 days.

14.6 The collection of samples

On days 0, 14, 28, 42, 60, 90, and 120, blood samples were taken from each group's animals. After killing four animals in each group on days 0, 90, and 120, samples of tissue, hair, and feces were taken.

14.6.1 Blood

In accordance with Brown19's protocol, pooled blood samples were taken from the tail veins of four rats from each group. One milliliter of the blood was kept in an EDTA-treated test tube for a hemogram, and 2 mL were immediately placed into pre-marked centrifuge glass test tubes in order to clot the blood and collect serum. Before being employed for biochemical parameters, the collected sera were stored at −20 °C.

14.6.2 Tissue

By following the conventional procedure and applying a larger dosage of ketamine, the rats were killed. Each rat's liver, kidney, heart, spleen, lung, and intestine were cut up and preserved for up to 48 h in 10% buffered formalin.

14.6.3 Hemogram

According to the accepted method20, hemoglobin, total RBC, total WBC, PCV, and differential count were all calculated.

14.7 Biochemical conditions

Measurements were made of serum AST and ALT activity21, BUN22, and CRT23.
Biochemical tissue (anti-oxidant status).
The activity of reduced glutathione (GSH), superoxide dismutase (SOD), MDA, and catalase was assessed in liver, kidney, and heart tissues.

14.8 Statistical evaluation

The measured parameters' values were presented as mean SE. Using a univariant general linear model with two methods, the mean values of several groups were compared. SPSS 10's ANOVA version of the program.

15. The findings and discussion

15.1 *B. acuminata* stem bark powder

It was observed that the levels of hemoglobin, total RBC, WBC, PCV, lymphocyte, and neutrophil counts did not change on the corresponding days for group

A0 animals, but the aforementioned values (groups A1, A4, & A5) significantly ($p < 0.05$) decreased until 90 days with respect to the "0" day value in the A1, A4, and A5 groups. On day 120, the values in group A1 substantially (p0.05) declined, but they rose in group A4 and group A5 treated with *Bauhinia acuminata* stem bark powder [41].

The powdered stem bark of *Bauhinia acuminata* may have diverse ameliorative effects on rats with chronic arsenicosis, and it is also quite inexpensive. So, it has a big opportunity to be employed as an affordable treatment in the future to treat human arsenic poisoning.

Here are some studies that may be relevant:

In a study published in the journal Food and Chemical Toxicology, researchers investigated the effects of *Bauhinia acuminata* stem bark extract on rats with liver damage induced by carbon tetrachloride. The study found that the extract improved liver function and reduced oxidative stress in the rats [42].

Another study published in the journal BMC Complementary and Alternative Medicine examined the effects of *Bauhinia acuminata* stem bark extract on diabetic rats. The study found that the extract reduced blood glucose levels and improved insulin sensitivity in the rats [43].

A third study published in the journal Pharmacognosy Research investigated the effects of *Bauhinia acuminata* stem bark extract on inflammation and pain in rats. The study found that the extract had anti-inflammatory and analgesic properties [44].

In a review article published in the journal Journal of Ethnopharmacology, the authors highlighted the traditional uses of *Bauhinia acuminata* L. in various regions of the world. The authors also discussed the potential pharmacological activities of the plant, including anti-inflammatory, antioxidant, and antimicrobial properties [45] .

It is important to note that while these studies may provide some insight into the potential effects of *Bauhinia acuminata* L. stem bark powder in rats, more research is needed to fully understand its differential effects. Additionally, it is important to consult with a healthcare professional before using any herbal supplements, as they may interact with medications or cause adverse effects.

15.2 *Bauhinia acuminata* stem bark extract

Baicalein (containing Flavonoids) concentration was calculated in *Bauhinia acuminata* ethanolic stem bark extract. The findings demonstrated that group G0 (control) total arsenic concentration in hair, feces, and essential organs did not change substantially over time in comparison to its "0" day value. Nevertheless, levels of arsenic were considerably higher on 90 and 120 days in group G1 (untreated control) than on the corresponding "0" day in the organs (lung, liver, kidney, heart, spleen, gut, and muscle), hair, and feces. On day 90, groups G4 and G5 likewise had a substantial (P0.05) rise in the amount of arsenic in their essential organs. One way that arsenic is eliminated is through the hair, but when BA stem bark extract is administered, arsenic is quickly removed from these organs. Exposure to either arsenite or arsenate causes an initial buildup in the liver, kidneys, lungs, and digestive system.

When compared to its value on day "0," the arsenite proportion in group G0's liver, hair, and feces did not change considerably over time in the same experiment. Nevertheless, group G1's arsenite percentage considerably rose on days 90 and 120, but group G4 and G5's arsenate fraction dramatically dropped on days 90 and 120. The organo-arsenic proportion was also considerably lower ($P < 0.05$) on day 90 in the liver, hair, and feces, but higher on day 120. Intoxicated human tissues showed relatively little accumulation of arsenate.

Arsenic's carbon metabolism and methylation can result in the formation of methylated arsenic metabolites such MAIII or trimethyl arsenic acid in both humans and animals.

The overall quantity of organo-arsenic species found in the rats in the current investigation was considerably less than the value found in hair and liver feces on day "0". In rats with arsenicosis, the concentration of arsenite was greater than that of arsenate, which was then, in comparison to the corresponding "0" days. This shows that inorganic arsenic is what causes toxicity. Nevertheless, animals of groups G4 and G5 treated with stem bark extract of BA at various dosage levels showed higher levels of organo arsenic and lower levels of both arsenite and arsenate [46].

Chronic exposure to arsenic can lead to various health issues, including skin lesions, respiratory problems, and cancer. Arsenicosis is a major public health problem in many parts of the world, particularly in South and Southeast Asia. Traditional medicinal plants have been widely used to treat arsenicosis. One such plant is *Bauhinia acuminata*, commonly known as dwarf white orchid tree.

Several studies have investigated the ameliorative effects of *Bauhinia acuminata* stem bark extract against chronic arsenicosis in rats. In one study, male Wistar rats were exposed to arsenic for 12 weeks, which resulted in significant biochemical and histological changes. Treatment with *Bauhinia acuminata* stem bark extract for 4 weeks significantly improved the biochemical parameters and prevented histological changes caused by arsenic exposure. The study concluded that *Bauhinia acuminata* stem bark extract has a protective effect against chronic arsenicosis in rats [47].

In another study, female Wistar rats were exposed to arsenic for 90 days, which resulted in significant oxidative stress and inflammation. Treatment with *Bauhinia acuminata* stem bark extract for 30 days significantly reduced oxidative stress and inflammation caused by arsenic exposure. The study suggested that the protective effect of *Bauhinia acuminata* stem bark extract may be due to its antioxidant and anti-inflammatory properties [48].

A third study investigated the protective effect of *Bauhinia acuminata* stem bark extract against arsenic-induced DNA damage in rats. Male Wistar rats were exposed to arsenic for 12 weeks, which resulted in significant DNA damage. Treatment with *Bauhinia acuminata* stem bark extract for 4 weeks significantly reduced the DNA damage caused by arsenic exposure. The study suggested that the protective effect of *Bauhinia acuminata* stem bark extract may be due to its ability to scavenge free radicals and prevent DNA damage [49].

In conclusion, the stem bark extract of *Bauhinia acuminata* has shown promising results in treating chronic arsenicosis in rats. The ameliorative effects of this plant may be due to its antioxidant, anti-inflammatory, and DNA protective properties. Further studies are needed to explore the potential therapeutic uses of this plant in treating arsenicosis in humans.

15.3 Leaf powder from *Glycosmis pentaphylla*

According to this study, animals in group A1 had considerably higher complete arsenic concentrations in their essential organs, hair, and feces on various days than did animals in group A0, the control group. After treatment with *G. pentaphylla* powder, the concentration of total arsenic was similarly considerably higher on day 90 in the animals in groups A2 and A3 compared to the corresponding "0" day, but it was lower on day 120 compared to day 90. The internal organ arsenic concentration was considerably greater in all arsenicosis rats. Among the group of A0 on various days, there was no discernible change in the proportion of arsenite found in the liver, hair, or feces. On the other hand, the arsenite percentage considerably increased ($p < 0.05$) in group on days 90 and 120 [50].

Glycosmis pentaphylla (Retz) is a plant commonly found in India and Southeast Asia, which has been traditionally used for various medicinal purposes, including the treatment of skin diseases and liver disorders. The leaves of this plant contain several bioactive compounds, including flavonoids, alkaloids, and phenolic acids, which have been shown to possess antioxidant, anti-inflammatory, and hepatoprotective properties.

A study conducted by Bhattacharjee et al. investigated the effects of DC leaf powder from *Glycosmis pentaphylla* on rats with chronic arsenicosis. The study involved administering arsenic to the rats through drinking water for 6 months to induce arsenicosis. The rats were then divided into four groups, with one group receiving only arsenic, and the other three groups receiving different doses of DC leaf powder from *Glycosmis pentaphylla* in addition to arsenic [51].

The results of the study showed that the rats treated with DC leaf powder from *Glycosmis pentaphylla* exhibited a significant improvement in various biochemical parameters, including liver function, oxidative stress, and inflammation, compared to the rats that received only arsenic. The researchers attributed these beneficial effects to the antioxidant and anti-inflammatory properties of the bioactive compounds present in the DC leaf powder.

Another study conducted by Singh et al. (2021) also investigated the hepatoprotective effects of *Glycosmis pentaphylla* against arsenic-induced liver damage in rats. The study found that treatment with the plant extract significantly reduced the levels of various liver enzymes and lipid peroxidation markers, indicating a protective effect against arsenic-induced liver damage [52].

In conclusion, DC leaf powder from *Glycosmis pentaphylla* appears to have beneficial effects on rats with chronic arsenicosis, particularly in improving liver function, reducing oxidative stress and inflammation. Further studies are needed to investigate the potential of this plant as a therapeutic agent for arsenicosis in humans.

15.4 *Glycosmis pentaphylla* leaf extract

Studies have been conducted to investigate the potential of GP leaf extract in mitigating the effects of chronic arsenic exposure in rats.

One study published in the Journal of Environmental Pathology, Toxicology, and Oncology investigated the effects of GP leaf extract on chronic arsenicosis-induced oxidative stress in rats. The study found that GP leaf extract significantly reduced oxidative stress markers in the liver and kidneys of rats exposed to arsenic. Additionally, GP leaf extract was found to improve the antioxidant status of the rats, indicating its potential in reducing the toxic effects of arsenic [53].

Another study published in the Journal of Ethnopharmacology evaluated the protective effects of GP leaf extract on chronic arsenicosis-induced neurotoxicity in rats. The study found that GP leaf extract significantly improved cognitive function and reduced oxidative stress in the brains of rats exposed to arsenic. The study also noted that GP leaf extract showed no signs of toxicity, indicating its safety for use as a potential treatment for chronic arsenicosis [54].

Overall, these studies suggest that GP leaf extract may have potential as a natural remedy for chronic arsenicosis, given its antioxidant and neuroprotective properties. However, further studies are needed to determine the safety and efficacy of GP leaf extract in humans.

16. Conclusion

In conclusion, arsenic is a toxic element that is naturally occurring in the earth's crust and can also be released into the environment through human activities such as mining, agriculture, and industrial processes. Exposure to arsenic can cause a range of health problems, including skin lesions, cancer, and neurological damage.

The sources of arsenic in the environment are numerous and varied, with contamination occurring in both natural and human-made settings. Arsenic can be found in soil, water, air, and food, and its presence can be difficult to detect.

The implications of arsenic contamination are significant and far-reaching. It poses a serious risk to human health, particularly in areas where access to safe drinking water is limited. It also affects agricultural productivity, ecosystem health, and economic development.

Remedies for arsenic contamination include a range of approaches, such as improving water treatment processes, promoting sustainable agricultural practices, and developing new technologies for detecting and removing arsenic from the environment. These efforts require a multi-disciplinary approach and cooperation between governments, industry, and communities.

Overall, addressing the issue of arsenic in the environment requires a concerted effort at the global, national, and local levels to reduce exposure and promote sustainable development practices. It is a complex issue that demands ongoing research, innovation, and collaboration to mitigate its harmful impacts.

Author details

Amartya De* and Nilanjana Roy
BCDA College of Pharmacy and Technology, Barasat, India

*Address all correspondence to: amartyap.col30@rediffmail.com

References

[1] Agency for Toxic Substances and Disease Registry. Toxicological Profile for Arsenic. Atlanta, GA: U.S. Department of Health and Human Services, Public Health Service; 2007

[2] Smedley PL, Kinniburgh DG. A review of the source, behaviour and distribution of arsenic in natural waters. Applied Geochemistry. 2002;**17**(5):517-568

[3] Bhattacharya P, Welch AH. Arsenic in groundwater: The deepening crisis in South and Southeast Asia. Environmental Science & Technology. 2009;**43**(17):651-657

[4] Ravenscroft P, Brammer H, Richards K. Arsenic Pollution: A Global Synthesis. United Kingdom: John Wiley & Sons; 2009

[5] Meharg AA, Hartley-Whitaker J. Arsenic uptake and metabolism in arsenic resistant and nonresistant plant species. New Phytologist. 2002;**154**(1):29-43

[6] Li RY, Ago Y, Liu WJ, Mitani N, Feldmann J, McGrath SP, et al. The rice aquaporin Lsi1 mediates uptake of methylated arsenic species. Plant Physiology. 2009;**150**(1):207-212

[7] Tripathi RD, Srivastava S, Mishra S, Singh N, Tuli R, Gupta DK, et al. Arsenic hazards: Strategies for tolerance and remediation by plants. Trends in Biotechnology. 2007;**25**(4):158-165

[8] Zhao FJ, McGrath SP, Meharg AA. Arsenic as a food chain contaminant: Mechanisms of plant uptake and metabolism and mitigation strategies. Annual Review of Plant Biology. 2010;**61**:535-559

[9] Meharg AA, Rahman MM. Arsenic contamination of Bangladesh paddy field soils: Implications for rice contribution to arsenic consumption. Environmental Science & Technology. 2013;**47**(7):2995-3001

[10] Zavala YJ, Gerhardt KE, Bhattacharya P, Telfeyan K, Ma LQ. Soil arsenic bioavailability and uptake by maize (Zea mays) influenced by rhizosphere phosphorus, copper, and iron. Environmental Pollution. 2008;**155**(3):502-509

[11] Zhang J, Cao Y, Bai J, Chen Z. Arsenic hyperaccumulation in Pteris vittata L.: A review. Ecotoxicology and Environmental Safety. 2017;**135**:157-163

[12] Dasgupta P, Liu Y, Dyke JV, Hwang SA. Association of arsenic exposure with smoking, alcohol, and caffeine consumption: A cross-sectional study from NHANES, 2003-2006. Environmental Health Perspectives. 2008;**116**(6):754-760

[13] World Health Organization. Arsenic. 2018. Retrieved from: https://www.who.int/news-room/fact-sheets/detail/arsenic

[14] National Institute of Environmental Health Sciences. Arsenic. 2019. Retrieved from: https://www.niehs.nih.gov/health/topics/agents/arsenic/index.cfm

[15] United States Environmental Protection Agency. Arsenic. 2022. Retrieved from: https://www.epa.gov/chemical-research/arsenic

[16] Karim MR, Hossain MB, Rahman MM, Sultana S. Arsenic contamination in groundwater and its effects on human health with particular reference to Bangladesh: A review.

Journal of Environmental Science and Health, Part A. 2019;**54**(6):503-516

[17] Rahman MA, Hasegawa H, Ueda K, Maki T. Removal of arsenic from groundwater by a floating macrophyte, water hyacinth (Eichhornia crassipes). Environmental Pollution. 2007;**147**(2):402-409

[18] Mohan D, Pittman CU Jr. Arsenic removal from water/wastewater using adsorbents—A critical review. Journal of Hazardous Materials. 2007;**142**(1-2):1-53

[19] Gupta DK et al. Arsenic detoxification potential of soil microbes and plants in contaminated environments: A review. Environmental Pollution. 2020;**263**:114509

[20] Ma LQ et al. Arsenic hyperaccumulation and its potential role in phytoremediation. Chinese Journal of Geochemistry. 2019;**38**(2):125-134

[21] Mandal BK, Suzuki KT. Arsenic round the world: A review. Talanta. 2002;**58**(1):201-235

[22] Mohanty S et al. Microbial bioremediation of arsenic-contaminated soil and water: Prospects and challenges. Environmental Science and Pollution Research. 2020;**27**(2):1094-1114

[23] Rascio N, Navari-Izzo F. Heavy metal hyperaccumulating plants: How and why do they do it? And what makes them so interesting? Plant Science. 2011;**180**(2):169-181

[24] Raskin I, Ensley BD, editors. Phytoremediation of Toxic Metals: Using Plants to Clean Up the Environment. New York: John Wiley & Sons; 2000

[25] Kumpiene J, Ore S, Renella G. Assessment of arsenic mobility in contaminated soils using sequential extraction procedures: A critical review. Environmental Pollution. 2014;**195**:1-21

[26] Meharg AA, Sun GX, Williams PN, Adomako E, Deacon C, Zhu YG, et al. Inorganic arsenic levels in rice grain: Origins, uptake mechanisms and toxicity. Environmental Science & Technology. 2008;**42**:7177-7186

[27] Lombi E, Zhao FJ, Dunham SJ, McGrath SP. Phytoremediation of heavy metal-contaminated soils: Natural hyperaccumulation versus chemically enhanced phytoextraction. Journal of Environmental Quality. 2001;**30**:1919-1926

[28] Freeman JL, Zhang LH, Marcus MA, Fakra S, McGrath SP, et al. Spatial imaging, speciation, and quantification of selenium in the hyperaccumulator plants Astragalus bisulcatus and Stanleya pinnata. Plant Physiology. 2006;**142**:124-134

[29] Agency for Toxic Substances and Disease Registry. Toxicological Profile for Arsenic (Update). US Department of Health and Human Services, Public Health Service; 2007

[30] Saha JC, Dikshit AK, Bandyopadhyay M. A review of arsenic contamination in India and its health effects. Environmental Health: A Global Access Science Source. 2012;**11**(1):11

[31] Das S, Dash HR, Chakraborty J. Microbial bioremediation of arsenic-contaminated environments: Recent advances and future prospects. Environmental Science and Pollution Research International. 1 Aug 2016;**23**(16):15951-15964

[32] Smedley PL, Kinniburgh DG. Arsenic in groundwater and the environment. Advances in Agronomy. 2005;**64**:141-185

[33] World Health Organization. Guidelines for Drinking-Water Quality.

Vol. 1. Gutenbarg: World Health Organization; 2011

[34] Islam F, Azizullah A, Khattak MN, Richter P, Häder DP. Bacterial degradation of arsenic in contaminated water using indigenous bacteria isolated from arsenic-contaminated site. Pakistan Journal of Zoology. 2011;**42**:567-573

[35] Journal of Microbiology and Biotechnology. 1 Dec 2015;**25**(12):2041-2048

[36] Cai L, Liu G, Rensing C, Wang G. Genes involved in arsenic transformation and resistance associated with the mobilization of arsenic in groundwater of the Hetao basin, Inner Mongolia. Scientific Reports. 10 Feb 2017;7:41023

[37] Smith AH, Lingas EO, Rahman M. Contamination of drinking-water by arsenic in Bangladesh: A public health emergency. Bulletin of the World Health Organization. 2000;**78**(9):1093-1103

[38] Ng JC, Wang X, Hu Q. Advances in arsenic pollution prevention and control: From traditional to novel approaches. Journal of Hazardous Materials (United States of America). 2019;**371**:643-651

[39] Bhattacharya P, Jacks G. Arsenic in groundwater: A global perspective with emphasis on the Asian scenario. Journal of Water Supply: Research and Technology-AQUA. 2017;**66**(7):429-446

[40] Ravenscroft P, Brammer H, Richards K. Arsenic Pollution: A Global Synthesis. Wiley-Blackwell; 2009

[41] Amartya D, Bandyopadhyay SK, Mandal TK. Differencial Effect of Bauhinia acuminata-L Stem Bark Extract on Detoxification of Arsenic in Rats. 2022. ISSN: 2320-2882

[42] Dua TK, Dewanjee S, Khanra R, Bhattacharya N, Gangopadhyay M. Protective effect of *Bauhinia acuminata* Linn. against carbon tetrachloride induced hepatotoxicity in rats. Food and Chemical Toxicology. 2011;**49**(1):78-86. DOI: 10.1016/j.fct.2010.09.011

[43] Prabhu KS, Lobo R, Shirwaikar A, Shirwaikar A. Bauhinia acuminata Linn.—A comprehensive review. Journal of Pharmacy Research. 2011;**4**(7):2103-2106

[44] Choudhary M, Kumar V, Malhotra H, Singh S, Kumar R, Kumari S. Anti-inflammatory and analgesic potential of *Bauhinia acuminata* L. bark. Pharmacognosy Research. 2014;**6**(1):64-70. DOI: 10.4103/0974-8490.122908

[45] Mishra S, Aeri V, Gaur PK. Anti-inflammatory and analgesic activities of Bauhinia acuminata. Journal of Ethnopharmacology. 2012;**142**(1):28-34. DOI: 10.1016/j.jep.2012.04.003

[46] De A, Bandyopadhyay SK, Mandal TK, Banerjee MC. Ameliorative Effects of *Bauhinia acuminata* stem bark extract against chronic arsenicosis in rats. Toxicology and Environmental Health Sciences. 2022:2349-5162

[47] Jahan S, Das S, Mandal S, et al. Protective effect of *Bauhinia acuminata* against arsenic-induced biochemical and histological changes in rats. Toxicology International. 2014;**21**(2):161-168

[48] Bhattacharjee P, Bhattacharya S, Dey D, et al. Protective role of *Bauhinia acuminata* against arsenic-induced oxidative stress and inflammation in female Wistar rats. Journal of Toxicology and Environmental Health. Part A. 2015;**78**(3):158-173

[49] Jahan S, Ali S, Mandal S, et al. Protective effect of *Bauhinia acuminata* against arsenic-induced DNA damage in Wistar rats. Indian Journal of Pharmacology. 2016;**48**(6):653-658

[50] De A, Bandyopadhyay SK, Mandal TK, Banerjee MC. Effects of *Glycosmis pentaphylla* (Retz) DC leaf powder against chronic Arsenicosis in rats. Wesleyan Journal of Research. Feb 2021;**14**(02)

[51] Bhattacharjee P, Paul S, Chatterjee A. Amelioration of arsenic induced oxidative stress in rat liver by *Glycosmis pentaphylla* (Retz) DC leaf powder. Journal of Ethnopharmacology. 2018;**212**:94-104

[52] Singh M, Patel R, Patel DK. *Glycosmis pentaphylla* (Retz.) DC. ameliorates arsenic induced liver damage in rats. Toxicology Reports. 2021;**8**:103-110

[53] Roy S, Ghosh A, Giri S. *Glycosmis pentaphylla* protects rat liver and kidney against chronic arsenic toxicity. Journal of Environmental Pathology, Toxicology, and Oncology. 2016;**35**(4):283-294

[54] Das S, Paul S, Khuda-Bukhsh AR. *Glycosmis pentaphylla* protects against arsenic-induced neurotoxicity in rats. Journal of Ethnopharmacology. 2012;**142**(1):68-78

Chapter 2

Arsenic Emission Control from Coal Combustion Flue Gas

Jiang Wu, Zhenzhen Guan and Yang Ling

Abstract

Arsenic is a highly toxic element in coal and one of the representative toxic trace metals emitted from coal-fired power plants, which is mainly converted into As_2O_3 vapor during the combustion process of coal. When absorbed by the body, arsenic can cause arsenic poisoning, which not only causes metabolic disorders and subsequent neurotoxicity in the body but also retards growth in young children. Arsenic is of increasing concern due to its bioaccumulation and potential carcinogenicity. This chapter describes the characteristics of arsenic emissions from coal-fired power plants and the various control technologies, including pre-, in-, and post-combustion control technologies. It also provides an outlook on future technological developments and provides theoretical guidance for controlling arsenic in flue gas.

Keywords: arsenic, technology, As_2O_3, coal-fired power plants, control

1. Introduction

As people's awareness of environmental preservation increases, the effects of heavy metal pollution on the environment have drawn considerable attention [1–3]. One of the heavy metal contaminants with high toxicity, bioaccumulation, and carcinogenicity is arsenic, a semi-volatile trace element [4, 5]. The majority of the As produced by coal combustion migrates into the flue gas as As_2O_3 vapor, some of which may interact with fly ash particles to form solids [6]. Dust collectors can catch the majority of the As that congeals in fly ash particles [7]. As poses a major health risk, when a small amount of it escapes into the atmosphere as tiny particles and vapors [8]. Additionally, in selective catalytic reduction (SCR) systems, arsenic vapor can interact with catalysts, causing these catalysts to malfunction [7]. In recent years, many researchers have investigated the type, content, and volatile properties of arsenic in coal. Pyrite, organic arsenic, and arsenate, which also contain arsenate, are the three primary forms of arsenic found in coal. Arsenate, on the other hand, is extremely stable and typically decomposes at rather high temperatures, whereas exchangeable and organically bound arsenic evaporates rapidly during the devolatilization of coal. And researchers have developed a number of techniques to control arsenic in the flue gas of coal-fired power plants. These include precombustion-, in-combustion-, and post-combustion technologies. The most common application today is the post-combustion removal of arsenic from coal-fired power plant flue gasses through the combined action of air pollution control equipment and sorbents. In addition, photocatalysis has shown great potential as a rapidly developing pollutant treatment method in recent years, with its low cost, simple process, nonpolluting nature, and recyclability. The removal of gaseous arsenic in flue gas is expected to be a new direction and many photocatalysts, including In_2S_3/g-C_3N_4, CoS/g-C_3N_4, $Bi_4O_5I_2$/g-C_3N_5, etc., have been developed by Wu et al. for investigating the removal of arsenic, mercury, and other pollutants from coal-fired flue gas. It provides theoretical guidance for the photocatalytic removal of arsenic from coal-fired flue gas.

2. Review of arsenic pollutants in coal-burning flue gas

2.1 Properties and hazards of arsenic

Elemental arsenic has a melting point of 817°C at atmospheres. When heated to 613°C, it sublimates directly without going through liquid. Arsenic can be divided into organic arsenic and inorganic arsenic and with speciations of trivalent arsenic (As^{3+}) and pentavalent arsenic (As^{5+}). The data show that inorganic arsenic is more toxic than organic arsenic, and As^{3+} is about 50 times more toxic than As^{5+}. Arsenic in coal mainly exists in three forms: pyrite, organic arsenic, and arsenate [9–11] among which arsenate includes the form of arsenate.

In the study of Guo et al. [12], arsenic in coal and coal-derived pyrolytic carbon can be divided into five forms: ion exchange type, carbonate binding type, iron and manganese oxide binding type, organic matter binding type, and residual arsenic in residue. Compared with raw coal, less arsenic is found in the forms bound to organic matter and iron and manganese oxides in coal coke produced at 1000°C, suggesting that these forms of arsenic are unstable during pyrolysis.

It is generally believed that during the devolatilization of coal, exchangeable and organically bound arsenic evaporates easily, while arsenic combined with iron-manganese oxides and pyrite is unstable and evaporates easily at temperatures below 1000°C, whereas arsenate is very stable and usually decomposes at relatively high temperatures. As a result, coals containing large amounts of unstable arsenic compounds have a higher volatilization rate at certain temperatures, while the arsenic in coal containing large amounts of stable arsenate is difficult to evaporate at low temperatures.

The coal-fired power stations, as the main source of coal consumption, may produce a large number of heavy metal pollutants, including arsenic [13], which has received increasing attention due to its toxicity, volatility, bioaccumulation in the environment, and potential carcinogenic properties. Coal-fired power plant ash treatment, such as the direct emissions of gaseous arsenic, is also harmful to the environment. When large quantities of fly ash are temporarily stored in stock or treated in fly ash landfills or lagoons [14], harmful elements in fly ash migrate or leach into the surrounding soil and groundwater due to the acid rain [15]. Bata et al. [16] reported that arsenic in fly ash was more readily leached than other heavy metals. However, Otero-Rey J R et al. [17] concluded that less than 20% of the total arsenic can be filtered out from fly ash. Therefore, effective control of toxic elements, such as arsenic, is recommended before using fly ash and gypsum as building materials.

2.2 Transport characteristics and emission paths of arsenic in coal-fired power plants

The arsenic emission during the combustion process has different forms of occurrence due to different migration paths in the power plant. There is solid phase arsenic fixed in bottom slag and gypsum, gas phase arsenic and re-emission with flue gas, and liquid phase arsenic of As^{3+} and As^{5+} in desulfurization wastewater and ash flushing water.

Arsenic mainly combines with fly ash in flue gas during migration. Senior et al. [18] pointed out that the conversion of arsenic to ash in coal is actually the migration of arsenic to components of different particle sizes in ash based on the formation and deposition of fine particles, that is. the distribution process of arsenic. Due to its high volatility, part of arsenic may still exist in the form of gaseous arsenic under the saturated vapor pressure formed at high temperature in the flue. In high-temperature flue, silicate melts in fly ash may dissolve arsenic and wrap it in fly ash as the temperature drops. In addition, the gaseous arsenic in the flue gas interacts with the calcium, aluminum, iron, and other metal elements in the fly ash to form arsenate. The flue gas arsenic in these two types of high-temperature flue is easy to be removed by flue gas desulfurization (FGD) device and dust removal device at the end of the flue, and part of the gaseous arsenic may condense on the surface of fly ash in the low-temperature flue. In the low-temperature flue, arsenic is physically adsorbed by the pore structure of fly ash, and gaseous arsenic condenses on the surface of fly ash due to the decrease of temperature [19]. Arsenic enrichment in fly ash tends to occur in succession with temperature changes during migration, and gaseous arsenic tends to be concentrated in submicron particles.

The arsenic enriched in fly ash migrates to low-temperature flue with the flue gas and is removed by wet FGD device, electrostatic precipitator (ESP) and wet electrostatic precipitator device, and fixed in desulfurization gypsum, fly ash, and other wastes.

At present, the control of heavy metal arsenic in coal-fired power plants mainly relies on the collaborative removal of ESP and Wet flue gas desulphurization (WFGD). However, ESP mainly removes trace elements in the form of PM, and it is not effective in removing arsenic from flue gas attached to submicron particles, which is easy to escape into the atmosphere. Although WFGD has a good effect on the removal of gaseous arsenic, the removed gaseous arsenic mainly enters the desulfurization wastewater in the form of PM and becomes arsenic in the liquid phase, which is more complex and difficult to remove. Only small amounts of arsenic-containing species are emitted in gaseous form. Therefore, the application of air pollution control devices (APCDs) can significantly affect the redistribution of arsenic in combustion byproducts and change the path of element to the environment.

2.3 Benefits of arsenic form transformation on arsenic removal from flue gas

Arsenic occurs in various forms in raw coal; but no matter what form exists in raw coal, any form of arsenic will come out in the form of As_2O_3 (g) under the high temperature of 1300°C $\sim$ 1600°C (1573 K $\sim$ 1873 K) in the furnace [6]. The volatilized arsenic will migrate to the bottom slag, gypsum, fly ash, wastewater, and other coal products with the process of combustion, and its form will change again. The results show that in standard oxidation system, arsenic mainly exists in the form of solid As_2O_3 when the temperature is below 750 K, and in the form of AsO (g), As_4O_6 (g), and As_2O_5 (s) when the temperature is between 750 K and 900 K. Only when the temperature is above 900 K, arsenic exists in a single form, AsO (g).

Arsenic in fly ash is mainly in the form of As^{5+}, while the concentration of As^{3+} is low, which is partly caused by the interaction of As_2O_3 (g) in the form of As^{3+} with CaO in fly ash to yield a stable As^{5+} compound [20, 21].

Arsenic speciations in coal-fired power plants are mainly As^{3+} and As^{5+}, among which gaseous arsenic is mainly As^{3+}. Many adsorbents have poor removal capability on $As3^{+}$ but have much better removal capability on As^{5+}. As^{3+} is converted to As^{5+} by preoxidation means, and then secondary treatment of arsenic adsorption is carried out, which can effectively reduce the toxicity of arsenic in coal combustion flue gas and the re-emission rate caused by thermal instability.

3. Arsenic emission control technologies for coal-fired power plants

3.1 Source control, removal, and mechanism of arsenic in flue gas

Existing control equipment is capable of removing arsenic compounds from flue gas, but it is not possible to achieve complete removal. This, coupled with the fact that total arsenic emissions from coal combustion remain a major source of arsenic in the atmosphere due to the huge volume of coal burned, poses a potential threat to the environment and human health. Recent years have seen the development of a number of control technologies to lower arsenic emissions from coal-fired power plants into the atmosphere. The stages of pre-, in-, and post-combustion are covered by these technologies.

Precombustion control technology is a series of treatments used to reduce elements, such as As contained in coal before it is used as an energy source. Arsenic is mainly present in coal in its inorganic-bound state, so the use of coal beneficiation technology to remove arsenic elements can reduce the amount of pollutants produced during combustion.

In order to separate coal from impurities, a type of coal processing technology known as coal preparation uses the physical and chemical differences between coal and impurities. Physicochemical coal preparation, chemical coal preparation, microbiological coal preparation, and physical coal preparation are the four categories under which coal preparation technologies can be categorized [22]. Four coal-washing plants in China were the subject of a study by Song et al. [23] on the washing removal of arsenic in raw coal. The findings showed that physical washing, with a removal effectiveness of more than 40%, could also successfully remove arsenic from raw coal. Flotation is a physicochemical coal-washing process that effectively reduces impurities, such as ash and sulfite in coal, and separates organic and inorganic minerals in coal through flotation agents. According to the results of Guo et al. study [24], coal contained between 73 and 83% of arsenic in a sulfide-bound condition. As a result, by removing sulfide from coal, sulfide ore flotation technology may also remove arsenic from coal. Akers et al. [25] explored the effect of coal washing on the removal of heavy metals from coal and found that conventional washing techniques could remove 47.1% of the arsenic, while advanced commercial washing techniques could be used to reduce even more. According to Wang et al. [26] and Zhou et al. [27] research, employing coal-washing technology under ideal circumstances, the average removal efficiencies of arsenic were 62.1% and 70.0%, respectively.

In recent years, with the continuous research on arsenic in coal, some scholars have found that the removal rate of arsenic from washed coal is closely related to its fugitive form in the coal. If the arsenic in the raw coal is mainly in the organic arsenic fugitive state or exists in fine minerals wrapped by organic matter, the phenomenon of arsenic enrichment in washed coal will occur, which makes it difficult to remove arsenic from the coal. The study of Wang et al. [28] showed that the occurrence of arsenic is complex at levels below 5.50 mg/kg, but overall, it is mostly in the organic state and is enriched in the washed coal during the washing process. According to Zhang et al. findings [29], bituminous coal could hardly be cleaned when arsenic was mostly organic in lignite. However, if the majority of the arsenic in bituminous coal was inorganic, it could be washed with a 74–100% washing efficiency. Therefore, precombustion washing technology has limitations in controlling the arsenic content in coal and is only applicable to the control of arsenic fugitive in the form of inorganic minerals. Currently. Precombustion stripping technology, although widely used in coal-fired power plants, is not the primary means, but only one of the auxiliary means.

3.2 Removal of arsenic from the furnace chamber

During coal combustion in high-temperature boiler furnace, arsenic is volatilized into the flue gas and is often enriched on fine fly ash particles, which makes it difficult for existing APCDs in coal combustion power stations to capture. Therefore, arsenic emissions can be reduced by inhibiting the volatilization of arsenic during coal combustion or by promoting the enrichment of arsenic on coarse particles, such as adding arsenic fixing agents and mixing coal combustion to convert arsenic from gaseous to granular state and from fine to coarse particles for subsequent removal treatment, thus improving the effectiveness of arsenic removal.

Currently commonly used additives are kaolinite, limestone, bauxite, etc. Gullett et al. [30] added several minerals to the furnace to control the emission of heavy metal elements from coal combustion and found that $Ca\ (OH)_2$, $CaCO_3$, and kaolinite all had significant inhibitory effects on the emission of arsenic during combustion. Zhao et al. [31] found that the arsenic mass concentration in the low-pressure impinger decreased from 0.25 mg/Nm^3 to 0.11 mg/Nm^3 after the addition of 3 wt% CaO to the raw coal compared with no arsenic fixing agent when burning southwest Guizhou coal at 1150–1400°C in a sinker, and CaO fixation effect was obvious. Xing et al. [1] investigated the ability of kaolinite to adsorb arsenic at elevated temperatures. Nearly, 40% of the trivalent arsenic [As(III)] was converted during the isolated arsenic adsorption into pentavalent arsenic [As(V)] and bonded to kaolinite, generating an As-O-Al structure. In this regard, kaolinite possesses a 200 μg g^{-1} arsenic adsorption capacity, of which 24% was found to be well crystalline Al bonded. Na-O-As-O-Al was created when O-Na groups bonded to As around the As-O-Al structure during co-adsorption, oxidizing 82% of As(III) to As(V), and attaching to the Al surface of kaolinite. Arsenic's adsorption capacity increased to 878 μg g^{-1}, with 56% of the arsenic bound to aluminum being well-crystallized.

Blended coal combustion is a clean coal combustion technology that can reduce pollutant emissions from coal and improve boiler performance by mixing two or more different types of coal for combustion. Zhao et al. [32] showed that when Heshan bituminous coal and Hollin River lignite were mixed in a 1:1 ratio, arsenic in the fine particulate matter was transferred to the coarse particulate matter, and arsenic in PM10 was reduced by 33% compared to Heshan bituminous coal after combustion. This indicates that some change in the elemental arsenic in the coal occurs during the combustion of the blended coal. Analysis of Huolinhe lignite revealed that this coal was mineral rich and that burning it in blends with other coals effectively reduced the melting temperature of the ash and also promoted reactions between elements such as iron and calcium and aluminosilicates, thereby inhibiting the production of fine particulate matter. However, some studies found that the volatility of arsenic in blended coals was higher than the weighted average of the arsenic volatility of the two original coals at different ratios, resulting in an increase in arsenic volatility. Liu et al. [33] found that high arsenic lignite blended with different bituminous coals. The proportion of arsenic volatilization was higher in all cases than in the single coal combustion due to the high volatile fraction of the lignite that could promote the burning of coal, which, in turn, promoted the emission of arsenic. There are a number of limitations to the current use of coal blending for the removal of arsenic from coal. Specifically, coal type, coal quality, and mineral content significantly affect the effectiveness of the blending method for arsenic removal, and the method is only applicable to a certain range of situations.

Arsenic removal can be achieved with additives or by blending different types of coal, and the process and equipment requirements are simple. Its arsenic removal performance varies widely depending on the kind and percentage of additives, etc., therefore much study is necessary despite the low investment and operation expenses.

3.3 Fixed removal of arsenic from tailpipe flue gas pollution control equipment

Post-combustion flue gas arsenic pollution control technology reduces arsenic emissions by using preexisting air pollution control equipment and sorbents to immobilize and remove arsenic from the flue gas. Most of the elemental arsenic is released as compounds during the high-temperature process of coal combustion. Those arsenic that are not captured and solidified by the arsenic fixing agent enter the high-temperature flue in gaseous form. Therefore, adsorbents can be sprayed into the flue to promote the conversion of gaseous arsenic to particulate arsenic, which can then be removed by air pollution control devices. Commonly used adsorbents include activated carbon, fly ash, and inorganic minerals. They have varying degrees of adsorption effectiveness and can be effective in removing gaseous arsenic from flue gasses.

The most frequently used adsorbent in the field of lowering flue gas pollution is activated carbon (AC), which has a well-developed pore structure and a wealth of application expertise. Marczak et al. [34] used commercial activated carbon to reduce the arsenic concentration in bituminous coal combustion flue gas from 146.2 $\mu g\ m^{-3}$ to 17.3 $\mu g\ m^{-3}$ with a removal efficiency of 88.2%. Wu et al. [35] synthesized a biomass-based porous carbon adsorbent using a combination of hydrothermal and CO_2 physical activation, which had an arsenic removal capacity of 1004 mg/g and showed significant resistance to SO_2 and HCl poisoning. The carbon-based adsorbent has a strong adsorption effect on gaseous arsenic in flue gas, but the selectivity is poor, it is easily influenced by flue gas components and temperature, and the application cost is relatively high, which makes it difficult in practical application.

During coal combustion, arsenic in the coal can evaporate and then condense on the fly ash surface as the temperature drops. Arsenic, selenium, and other trace elements were the focus of Bartoňová et al.'s [36] study on the impact of unburned carbon in fly ash on the capture of 12 trace elements. The findings also showed that fly ash's carbonaceous matter had the ability to retain arsenic, and there was a correlation between the amount of trace elements and the amount of unburned carbon. Díaz-Somoano [37] investigated the adsorption process of fly ash for gaseous arsenic at 550°

C and 750°C, and the results showed that fly ash has good removal ability for arsenic, and temperature increase can promote the adsorption and capture ability of fly ash. Li et al. [38] found that at low temperatures (450°C) fly ash could capture gaseous arsenic by physical and chemical adsorption, while at high temperatures (900°C), it mainly captured gaseous arsenic by chemical adsorption.

Through the interaction between heavy metals and inorganic minerals, mineral sorbents have been developed for the progressive adsorption and capture of gaseous arsenic elements in flue gasses. Yu et al. [39] investigated the adsorption characteristics of mineral oxides on As_2O_3 in a two-stage reactor. The results showed that CaO was an excellent adsorbent for As_2O_3. In the range of 300–900°C, the adsorption of As_2O_3 by CaO, MgO, and Al_2O_3 became better as the temperature increased. At 300–700°C, the order of As_2O_3 adsorption was CaO > Fe_2O_3 > Na_2O > MgO > Al_2O_3 > SiO_2, and SiO_2 hardly adsorbed any As_2O_3.

Electrostatic precipitators are capable of removing particulate matter from flue gas using high-pressure electrodes, and as some of the arsenic is enriched in particulate matter, it can also be effective in synergistically removing the heavy metal arsenic. However, it is generally accepted that arsenic in particulate form represents a relatively small proportion of the total arsenic emissions from coal combustion and that electrostatic precipitators have a relatively low removal efficiency for submicron fine particulate matter, so the ability of electrostatic precipitators to remove arsenic is limited. Bag filters are more efficient at collecting arsenic than electrostatic precipitators. They use filter media to filter the dust out of the flue gas and are relatively unaffected by the characteristics of the dust, so the removal efficiency is higher. However, this method can have a greater impact on the resistance of the flue gas. In wet conditions, the bags can be glued and need to be replaced frequently. Wet flue gas desulphurization (WFGD) is currently the most widely used wet FGD method in China, and it removes some of the arsenic along with the desulphurization. In a study by Patricia Córdoba et al. [40], the average concentrations of both gaseous and particulate arsenic at the outlet of the wet flue gas desulphurization (WFGD) system were lower than at the inlet, and the arsenic concentrations in both the limestone slurry and the gypsum slurry were higher than the initial process water, indicating that arsenic can be fixed in the gypsum with the help of limestone and is also present in the desulphurization wastewater and that the system also acts as a synergistic arsenic removal. However, gypsum, a byproduct of wet FGD, will mix with arsenic to generate stable calcium arsenate, which will exacerbate the scaling issue with the equipment and have some bad impacts on human health, making it difficult to use gypsum as a resource. The addition of a wet electrostatic precipitator (WESP) allows further removal of particulate matter and heavy metals before the flue gasses are discharged to the environment. Within the WESP, a high-water vapor environment can facilitate the removal of PM2.5, acid mist, and tiny droplets. In the study by Wang et al. [41], their findings that arsenic is more likely to be enriched in tiny particles, hence removing small particles will help remove arsenic more effectively. At the WESP's inlet, arsenic concentrations were 0.130 $\mu g\ m^{-3}$, while at the exit, they were 0.098 $\mu g\ m^{-3}$. The WESP obtained a 24.6% As removal efficiency. The arsenic removal efficiency of the WESP after retrofitting it with a high-frequency power ESP and a low-temperature coal saver was 99.93%. This also demonstrates the improved arsenic capture capacity of the ultralow emission retrofit in the CFPP.

In conclusion, pollution control equipment is also an effective means to prevent the release of arsenic pollutants into the environment, as reflected in the synergistic removal of particulate matter, but the existing dust removal equipment is more difficult to capture sub-fine particles. There is still room for improvement.

4. Research on arsenic adsorption

4.1 Removal of arsenic by fly ash sorbents

Due to its stability and low production cost, fly ash as a byproduct of power plants has the potential to become an adsorbent. During coal combustion and condensation, a large proportion of gaseous arsenic in flue gas migrates to fly ash. This means that fly ash can be used as an adsorption material to remove trace elements.

In order to explore the adsorption capacity of fly ash for arsenic, López-Antón et al. [42] selected two fly ash samples as adsorbents in a fixed bed reactor to carry out a series of adsorption experiments. Two samples were taken from a fluidized bed combustion plant that burns a mixture of coal waste and a pulverized coal power plant that burns high-grade coal. Adsorption experiments at 120°C showed that the maximum retention capacity of fly ash for gaseous arsenic was 5.3 mg g^{-1}, and the corresponding maximum adsorption efficiency was less than 20%. Díaz-Somoano et al. [43] suggested that calcium and iron in fly ash play a key role in arsenic adsorption by measuring the content of calcium, iron, and arsenic. Since high temperature may cause the re-carbonization of unburned carbon in fly ash, resulting in the change of fly ash adsorbent, and 120°C is close to the temperature of ESP inlet, which is convenient for the injection of fly ash adsorbent, they studied the adsorption of gaseous arsenic by fly ash at 120°C. However, gaseous arsenic tends to condense on the tube wall at this temperature, increasing the possibility of major errors in such lab-scale adsorption experiments.

In recent years, further research has been conducted on the reaction mechanism of fly ash adsorption of gaseous arsenic based on practical applications in coal-fired power plants. Wang et al. [44] conducted ash spraying test at the outlet of SCR device of 1000 MW power plant in order to study the collaborative removal of trace elements (TEs) such as arsenic, selenium, and lead in flue gas. The results showed that the fume-modified fly ash could generally improve the TEs removal efficiency of electrostatic precipitator and flue gas desulphurization unit. Li et al. [38] studied the adsorption and removal of gaseous As_2O_3 by coal-fired fly ash through experiments and found that fly ash samples collected from different power plants could capture As_2O_3 to a certain extent, suggesting that this effect is universal. At low temperature, it is dominated by the synergistic effect of physical adsorption and chemisorption, and at high temperature, chemisorption provided by effective mineral components in fly ash is the main reaction mechanism.

4.2 Removal of arsenic by metal oxides

Studies have revealed a strong linear association between the amount of arsenic enriched in coal combustion fly ash and the active metallic mineral composition. As a result, oxides corresponding to these metal mineral compositions (e.g., Fe_2O_3, CaO, and γ-Al_2O_3) have been extensively investigated for the removal of gaseous arsenic. Also, metal oxides are gaining importance due to their lower cost, wide availability, and superior effectiveness. The adsorption of gaseous phase As_2O_3 on CaO, Fe_2O_3, and Al_2O_3 was examined by Zhang et al. [45, 46], and the adsorption of gaseous phase arsenic on CaO and Fe_2O_3 was mainly chemisorbed at 600–900°C. With rising temperature, effective adsorption is reduced. The best adsorbent for arsenic is iron oxide, followed by calcium oxide and aluminum oxide. Zhao et al. [47] studied how the composition of the flue gas and the adsorption temperature affected the removal of As_2O_3 by Fe_2O_3, CaO, and γ-Al_2O_3. The results showed that the removal capacity of As_2O_3 met $Fe_2O_3 >$ CaO $> \gamma$-Al_2O_3 in the range of 300–900°C. At 500°C, Fe_2O_3 had the maximum As_2O_3 removal capacity (adsorption capacity: 158.4 μg g^{-1}; removal efficiency: 84.68%). Arsenic was also more easily converted to high-valent arsenic (As (V)) on the surfaces of Fe_2O_3, CaO, and γ-Al_2O_3 by raising the adsorption temperature. Low SO_2 concentrations would prevent the three metal oxides from effectively removing As_2O_3 by competing with it for adsorption on their surfaces. The removal of Fe_2O_3, CaO, and γ-Al_2O_3 by As_2O_3 was significantly inhibited by the presence of NO but was unaffected by variations in NO concentration. Furthermore, calcium-based compounds are frequently utilized as sorbents because they are inexpensive and nontoxic. During capture, the adsorbent not only physically absorbs As_2O_3(g) but also chemisorbs, converting As(III) to the less toxic As(V). Li et al. [48] synthesized CeO_2/CaO adsorbent by sol-gel method and improved its sintering resistance and capture efficiency by doping CaO with CeO_2. The SEM images of the samples are shown in **Figure 1** [48].

As demonstrated in **Figure 1(b)**, when untreated CaO is compared to spent CaO at 900°C, untreated CaO has very small, more uniformly distributed particles. However, a major aggregation phenomenon is clearly observed in the spent sorbent following adsorption at temperatures below 900°C, resulting in sorbent sintering and a decrease in accessible active sites. In contrast, as demonstrated in **Figure 1(d)**, CeO_2/CaO did not change much before and after the reaction, implying that CeO_2 doping increased its sintering resistance.

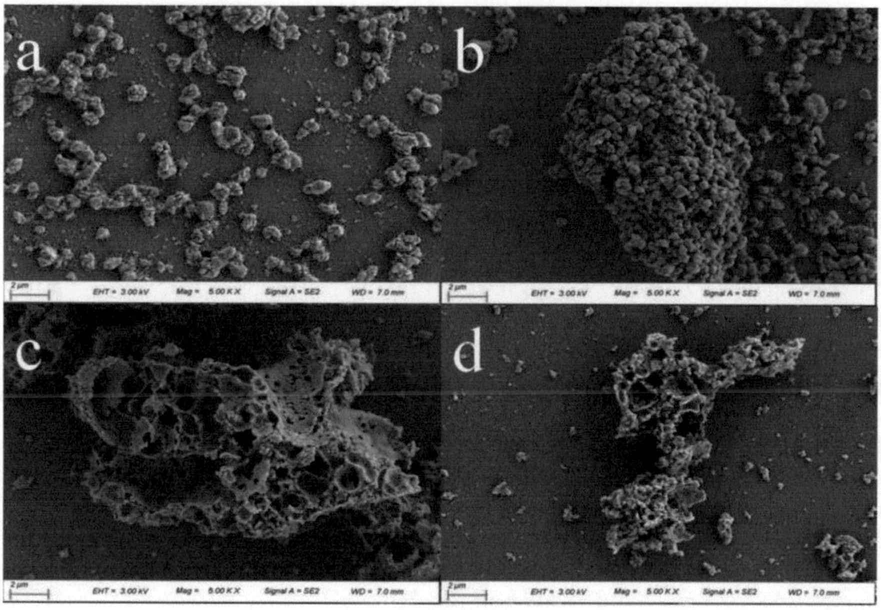

Figure 1.
SEM photograph of (a) untreated CaO, (b) CaO obtained at 900°C, (c) untreated CeO_2/CaO, and (d) CeO_2/CaO obtained at 900°C [48].

4.3 Fabrication of Al_2O_3/CaO with anti-sintering for efficient removal of As_2O_3

Due to its low cost and stable chemical properties, calcium oxide (CaO) has been extensively used as a high-temperature sorbent in the field of heavy metal removal. However, CaO tends to sinter at high temperatures, which degrades its adsorption performance. Using citric acid and calcine, a sintering-resistant Al_2O_3/CaO sorbent was synthesized, which was then applied in an innovative manner to the adsorption of As_2O_3 by combining experiment and theoretical analysis [46].

Based on the experimental findings, the potential internal mechanism of As_2O_3 adsorption on As_2O_3/CaO can be investigated further. CaO reacted with Al_2O_3 at a high temperature, producing $Ca_3As_2O_3$, a support framework with high thermal resistance and a stable structure. Al_2O_3 can provide a large quantity of specific surface area and pore structure, which can facilitate the physical adsorption process, despite the fact that purified Al_2O_3 hardly reacts with As_2O_3. SEM can, therefore, reveal the unique surface morphology of As_2O_3/CaO. Al loses electrons in the presence of As_2O_3 to form Al^{3+}, while CaO and As_2O_3 acquire electrons to form AlO_4 tetrahedral structure by combining with Al^{3+}. Population analysis and density of states reveal that Al becomes the principal active site of As_2O_3 adsorption and functions as the electron transport medium. In the meantime, the addition of Al can facilitate the separation of lattice oxygen, which may oxidize As_2O_3 and form a covalent bond on the adsorption surface. Therefore, Al_2O_3/CaO has a significantly greater absorption capacity than U-CaO.

4.4 Adsorption of arsenic by a-Fe_2O_3 adsorbent

The addition of an adsorbent to the combustion furnace can inhibit the volatilization of arsenic. Iron-based catalysts have been widely studied because they are widely available, inexpensive, harmless and have a large specific surface area. It is a technology that shows promise for removing arsenic from coal-fired flue gas. Among other things, iron-based adsorbents can remove elemental arsenic from flue gas by chemical reaction with arsenic or by physical adsorption; the surface lattice oxygen of Fe_2O_3 can oxidize $As_2O_3(g)$ to $As_2O_5(s)$ and form iron arsenate ($FeAsO_4$) [49].

The results of the DFT study by Zhang et al. [50] also showed that As_2O_3 (g) could form eight stable adsorption structures on the (001) surface of Fe_2O_3 with a minimum adsorption energy of 275.52 kJ mol^{-1}. One-step hydrothermal preparation was used by Zhao et al. [51] to create a tiny, spherical α-Fe_2O_3 adsorbent, and the effects of reaction temperature, nitric oxide (NO), oxygen (O_2), nitric dioxide (NO), sulfur dioxide (SO_2), and inlet arsenic concentration on the adsorption performance were investigated. The effect of its temperature on the capture of $As_2O_3(g)$ by α-Fe_2O_3 is shown in **Figure 2** [51].

The experimental findings in **Figure 2(a)** demonstrated that the tiny spherical α-Fe_2O_3 had a higher adsorption capacity for As_2O_3 than conventional Fe_2O_3. According to **Figure 2(b)**, temperature had a substantial impact on the adsorption capacity of α-Fe_2O_3. The adsorption capacity rose significantly when the temperature was raised from 500 to 600°C. However, the adsorption capability gradually declined after the temperature surpassed 600°C. Before and after the reaction, SEM of α-Fe_2O_3 was conducted; the findings are displayed in **Figure 3** [51].

Pure α-Fe_2O_3 develops a tetragonal morphology with a regular distribution on the spherical surface, as shown in **Figure 3(a)**, and it has an excellent spherical microstructure. Theoretically, compared to a smooth spherical surface, this form has more

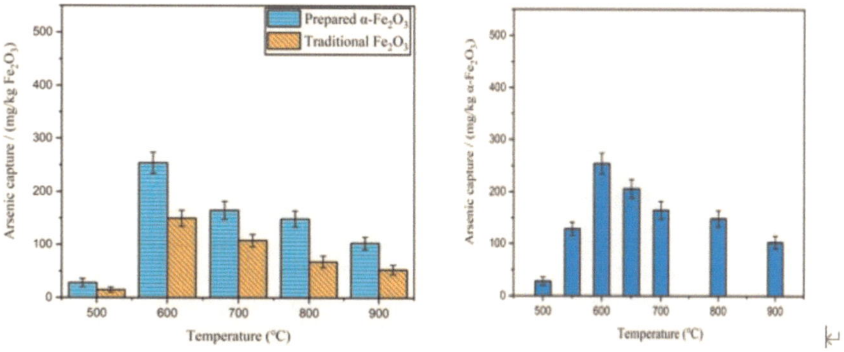

Figure 2.
(a) Arsenic capture by traditional Fe_2O_3 and the as-prepared α-Fe_2O_3 from 500 to 900°C. (b) Arsenic capture by α-Fe_2O_3 at temperatures from 500 to 900°C [51].

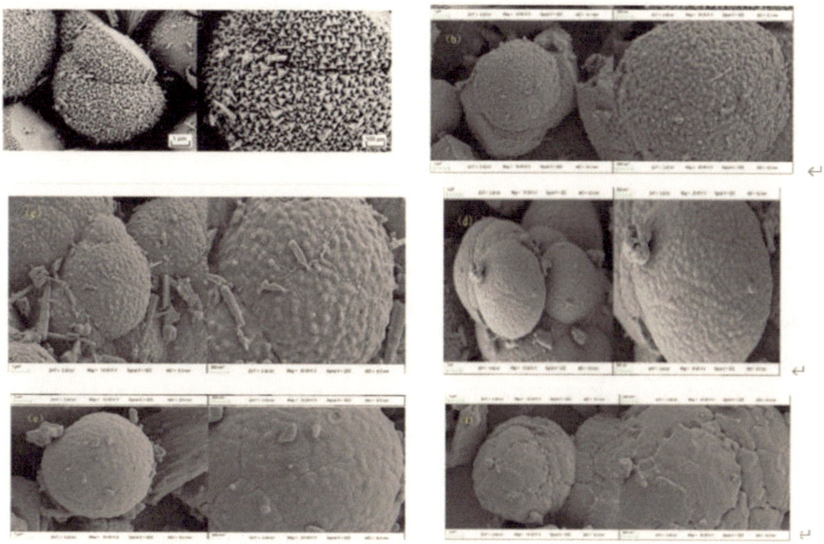

Figure 3.
SEM images of α-Fe_2O_3 before adsorption (a) and after adsorption at 500°C (b), 600°C (c), 700°C (d), 800°C (e), and 900°C (f) [51].

space for the physical adsorption of arsenic on the α-Fe_2O_3 surface due to its bigger specific surface area and adsorption active sites. The tetragonal angular morphology vanished and was replaced by a gradually smooth spherical shape when α-Fe_2O_3 absorbed arsenic in the experiment from 500 to 700°C (**Figure 3(b–d)**). This might be the outcome of the combined actions of calcination at high temperatures and arsenic adsorption. As the temperature increased from 800 to 900°C, it became more apparent that α-Fe_2O_3 was sintering and that there were less and fewer active sites available for adsorption (**Figure 3(e, f)**). Chemical adsorption was the key factor in the capture of arsenic.

In conclusion, the adsorption performance of α-Fe_2O_3 is temperature dependent. Due to the chemisorption mechanism, as the temperature rises from 500 to 600°C, its adsorption capability steadily increases. Its adsorption ability begins to decline as the temperature hits 600°C as a result of sintering. 600°C is the ideal adsorption temperature.

4.5 Catalytic adsorption of arsenic in the liquid phase

4.5.1 Preparation of CuO/TiO_2 adsorbent

CuO/TiO_2 adsorbent can be prepared by impregnation-thermal decomposition method. Weigh appropriate amount of Cu $(NO_3)_2{\cdot}3H_2O$ and pour it into a beaker, add 20 mL of deionized water, and stir until Cu $(NO_3)_2{\cdot}3H_2O$ is completely dissolved. 4.4 g anatase TiO_2 was added, heated and stirred to get a uniform slurry, and then dried in a blast dryer at 80°C for 12 h to get a light blue solid. The samples were crushed and heated in a muffle oven at 5°C/min to 500°C and baked at this temperature for 4 h. TiO_2 adsorbent with different CuO loads can be prepared successfully by taking out solids and grinding them carefully. Conditions to control the load: m (CuO): m (TiO_2) = 0.00:1, 0.01:1, 0.10:1, and 0.20:1 (abbreviated as CT0, CT1, CT10, and CT20). For comparison, ordinary CuO without TiO_2 is also obtained by calcination of Cu $(NO_3)_2{\cdot}3H_2O$.

4.5.2 Removal of arsenic in liquid phase by CuO/TiO_2

Liquid-phase arsenic removal experiments are conducted in a laboratory-scale system. The appropriate amount of sodium arsenate solution was slowly poured into the adsorption reactor to ensure that the concentration of As(III) was 25 mg/L, and then the appropriate amount of adsorbent (CuO/TiO_2, TiO_2, and CuO) was added into the solution. Then, the magnetic agitator was started, and the adsorption was carried out dynamically at room temperature (25°C) at a rate of 300 r/min for 4 h. The constant temperature tank is connected to the adsorption reactor to ensure that the temperature will not interfere with the experimental results. At the end of the experiment, a syringe equipped with Millipore membrane filter (0.22 μm) was inserted into the extraction port to extract 4.0 mL of the reaction suspension. The filtered supernatant could be directly used in ICP-OES to determine the concentration of As in the supernatant without further treatment. The removal efficiency of arsenic at adsorption equilibrium is calculated as follows:

$$\eta = \frac{C_0 - C_e}{C_0} \times 100\% \tag{1}$$

where C_0 and Ce are As concentration (mg/L) of solution before and after adsorption, respectively.

4.5.3 Effect of different CuO doping ratio on removal performance of arsenic in liquid phase

Different doping ratios of CuO have great influence on the crystallinity and morphology of CuO/TiO_2 adsorbent and also affect the removal effect of CuO/TiO_2 adsorbent on arsenic. Arsenic removal efficiency of CT0 (TiO_2), CT1, CT10, CT20, and CuO samples after data processing is shown in **Figure 4**. CT0 is untreated anatase TiO_2, and its removal efficiency for arsenic reaches 73.89%, indicating that pure TiO_2 is indeed a material with good adsorption performance for arsenic. For CuO/TiO_2

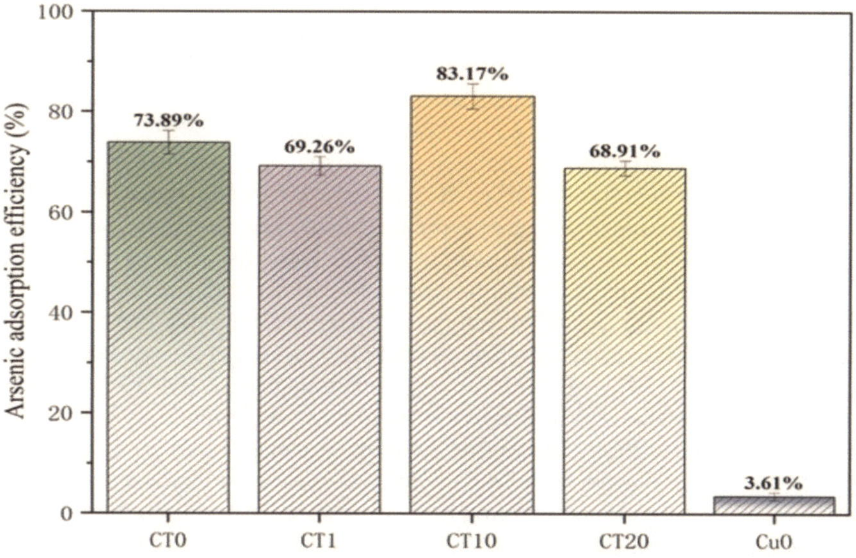

Figure 4.
Comparison of arsenic removal capacity of different doping ratios of CuO.

samples, the doping ratio of CuO is the key factor to determine the removal rate of arsenic. As can be seen from the figure, with the increase of CuO content, the removal efficiency of arsenic first increased and then decreased, and the highest removal efficiency of arsenic was CT10, with an efficiency of 83.17%. In the ascending stage of CT1 to CT10, with the increase of doping ratio, the removal effect of adsorbent on arsenic also increases. This is due to the doping of Cu element, which increases the original specific surface area of TiO_2 and exposes more active sites. The reason why the removal efficiency of CT20 decreases is that with the continuous doping of CuO particles, the proportion of CuO increases, and the CuO is excessively attached to the surface of TiO_2 crystal, occupying the effective active site on the surface of TiO_2. In addition, the agglomeration of CT20 reduces the specific surface area of the adsorbent, resulting in the decrease of arsenic removal performance [52]. The removal efficiency of the CuO reference group was 3.61%, indicating that CuO alone has no affinity for arsenic and is difficult to be used as an adsorbent for arsenic removal alone. Therefore, appropriate proportion of CuO doping (CT10) can improve the adsorption performance of TiO_2 for arsenic.

5. Theoretical study on arsenic adsorption

The composition of flue gas arsenic typically includes both elemental arsenic and various forms of oxidized arsenic, including AsO, AsO_2, As_2O_3, and As_2O_5. The compound As_2O_3, which comprises trivalent arsenic (As^{3+}), exhibits stable thermodynamic properties, presents a challenging dissociation process, and poses significant toxicity risks, thereby causing severe environmental and human health impacts. In addition, the presence of gaseous arsenic has the potential to impede the functionality of selective catalytic reduction (SCR) mechanisms, thereby inducing arsenic poisoning of SCR catalysts, which could ultimately diminish the financial advantages of power plants. Theoretical calculations are necessary to examine the mechanism of gaseous arsenic in fly ash, with the aim of comprehensively understanding the pattern of transformation and migration of flue gas arsenic to a particle state [53].

In recent years, density functional theory (DFT)-based quantum chemical computing has been increasingly utilized in pollution control research.

The research conducted in this study involved the development of an unburned carbon model for fly ash. Additionally, the researchers performed density functional theory (DFT) calculations to investigate the impact of surface defects, surface active chlorine groups, and an acidic gas atmosphere on the adsorption of elemental mercury. The theoretical findings were found to be consistent with the experimental results, leading to a comprehensive analysis of the aforementioned factors. This information is documented in Ref. [54]. A separate research endeavor was conducted to examine the impact of oxygen-containing functional groups (OCFGs) on the adsorption process of elemental mercury from unburned carbon in fly ash. The study revealed that OCFGs exhibited a notable ability to enhance the reactivity of neighboring carbon atoms. Furthermore, the quantity of carbon and hydrogen atoms present in the functional groups was found to have a substantial influence on the adsorption properties of mercury. In contrast, conventional density functional theory (DFT) calculations are typically performed assuming a temperature of absolute zero and an absolute vacuum. This poses challenges in establishing a direct correlation between the obtained results and the real-world operating conditions of power plants. Consequently, this can lead to distortions or even hinder the formation of conclusive findings [55]. Hence, the inclusion of the fundamental principles of thermodynamics becomes imperative in order to extrapolate the outcomes of density functional theory (DFT) calculations to a particular temperature range and effectively align with experimental or industrial circumstances. Nevertheless, there is a scarcity of research examining the dynamics of the interaction between gaseous arsenic and unburned carbon in fly ash. Additionally, a comprehensive investigation into the morphological characteristics of arsenic at different temperature intervals is lacking. In this section, the utilization of the DFT coupled thermodynamics method was employed to conduct a comprehensive investigation into the conditions necessary for the existence of flue gas arsenic, as well as the preferred structure it adopts, within the temperature range relevant to power plants. The adsorption behavior of various gaseous arsenic species was investigated. The objective of this study was to examine the structure of preferential adsorption and the key factors that influence the adsorption of arsenic on carbon-based carriers. The findings of this study are expected to contribute to the mitigation of arsenic pollution in coal-fired power plants [56, 57].

5.1 The theoretical model of arsenic adsorption

The theoretical framework relies on density functional theory (DFT) calculations, which are executed using the preexisting DMol 3 code programs [57]. The characterization of the system's exchange-correlation functional was conducted using the Perdew-Burke-Ernzerhof (PBE) version of the generalized gradient approximation (GGA) functional [58]. The convergence tolerance was defined in the study as the maximum energy change of 2.0×10^{-5} Ha, the maximum force of 0.004 Ha/Å, and the maximum displacement of 0.005 Å [59]. The utilization of effective core potentials was employed in order to accurately describe the behavior of the core electrons. Additionally, the calculation was performed using the double numeric with polarization (DNP) basis set. The energy, population analysis, and density of states were subsequently calculated based on the geometrically optimized model. To evaluate the structural stability, the adsorption energy can be determined using the following formula.

$$E_{ad} = E_{AB} - E_A - E_B \tag{2}$$

where E_{ad} represents the adsorption energy (eV). E_{AB} is the total energy (eV) of the stable structure where B molecule adsorbed on the A molecular surface. E_A and E_B represent the total energy (eV) of the isolated A surface and B molecule, respectively.

To bring the results of DFT calculations closer to the actual conditions of an experiment or industrial application, thermodynamic methods must be used to extrapolate its conclusions. As one of the state parameters of a thermodynamic system, the Gibbs free energy can be used to determine the direction of a chemical reaction, as described below:

$$G = H - T \times S = U + P \times V - T \times S \tag{3}$$

where G is the free energy of Gibbs (J/mol). Enthalpy (J mol^{-1}), temperature (K), entropy (J/mol K), internal energy (J mol^{-1}), pressure (Pa), and molar volume (m^3 mol^{-1}) are represented by *H, T, S, U, P,* and *V*, respectively [59]. However, the contribution of zero-point vibration is disregarded. Thus, the Helmholtz free energy can be calculated using (Eq. (4)), which accomplishes the coupling with the DFT results:

$$F = E_{DFT} + F_{vib}F = E_{DFT} + F_{vib} \tag{4}$$

where F is the Helmholtz free energy (J mol^{-1}). E_{DFT} is the total energy from the DFT calculation (J mol^{-1}). While F_{vib} takes into account the contribution of vibration to the system energy (J mol^{-1}), which can be obtained by (Eq. (5)):

$$F_{vib} = E_{ZPVE} - T \times S \tag{5}$$

where E_{ZPVE} is the zero-point vibrational energy at the absolute zero (J mol^{-1}). Next, by using the definition of the Helmholtz free energy (Eq. (6)), the Gibbs free energy can be determined by (Eq. (7)).

$$F = U - T \times S \tag{6}$$

$$G = E_{DFT} + E_{ZPVE} + P \times V - T \times S \tag{7}$$

Finally, based on the ideal gas equation of state, the Gibbs free energy per unit mole of the system is obtained by (Eq. (8)).

$$G = E_{DFT} + E_{ZPVE} + R \times T - T \times S \tag{8}$$

where R is the universal gas constant, which is equal to about 8.314 J/mol $\times$ K in the international system of units. In addition, the change in Gibbs free energy for a reaction can be determined by (Eq. (9)).

$$\Delta G = GP - GR \tag{9}$$

where *GP* and *GR* represent the Gibbs free energy of the reactant and product (J mol^{-1}), respectively. (For the convenience of description, the unit of kcal mol^{-1} is adopted in the following discussion, and its conversion relation is 1 kcal mol^{-1} = 4184 J mol^{-1}).

5.2 Mechanism study of arsenic adsorption

The present study focuses on a comprehensive examination of the various potential forms of arsenic in flue gas, including As, AsO, AsO_2, As_2O_3 and As_2O_5. The polar molecular structure of AsO_2 arises from the positioning of two oxygen atoms on opposite sides of the arsenic atom. This configuration results in an As-O bond length of 1.664 Å and an O-As-O angle of 128.239 [60].

As illustrated in **Figure 5**, there are six distinct forms of gaseous As_2O_3 denoted as As_2O_3-x (where x ranges from 1 to 6). The As_2O_3-1 structure exhibits a notable superiority compared to the remaining six structures under conditions of low temperatures. However, with increasing temperature, the differentiation between As_2O_3-1 and As_2O_3-2 (or As_2O_3-4) gradually decreases. The stability of the As_2O_3-4 structures, exhibiting a chain-like arrangement, is observed to reach its maximum at a temperature of 900 Kelvin. The stability of As_2O_3-2, which bears a resemblance to a horn, is greater than that of As_2O_3-1 when exposed to temperatures exceeding 900 K [61, 62]. However, it is surpassed in stability only by As_2O_4-4. In earlier studies, it was widely accepted that the presence of As^{3+} in coal-fired flue gas predominantly manifested in a trigonal bipyramidal configuration, such as As_2O-1, characterized by spatial symmetry and structural stability. At low temperatures, As_2O_3-1 is the dominant species. However, when the temperature surpasses 900 K, the likelihood of As_2O_3-2 and As_2O_3-4 being present becomes greater than that of As_2O_3-1. This observation implies that under elevated temperatures, the molecular bonds within the trigonal bipyramidal structure of As_2O_3-1, which are typically stable, will undergo dissociation, resulting in the formation of alternative structures (As_2O_3-2 and As_2O_4) that are not present under lower temperature conditions. Furthermore, the

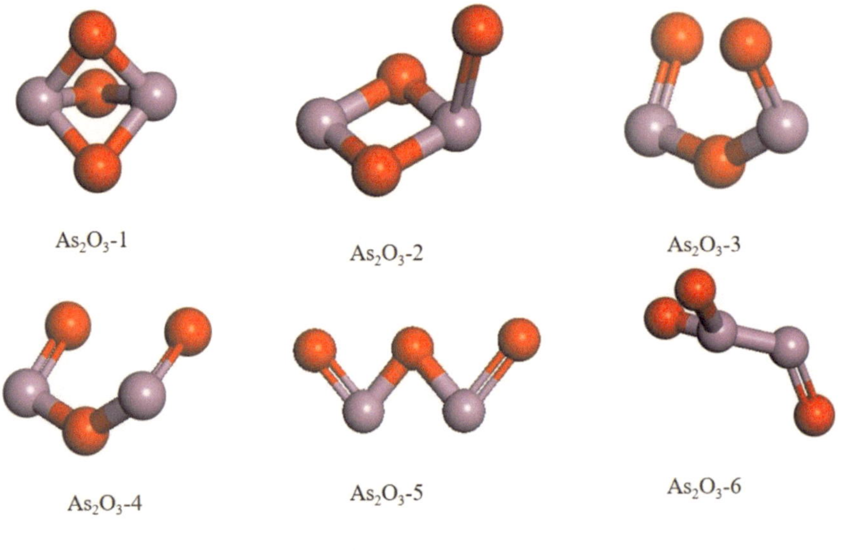

Figure 5.
Schematic diagram of the candidate As_2O_3 [53].

importance of the DFT combined thermodynamics approach in analyzing the morphology of arsenic in coal-fired flue gas under different operational temperatures is elucidated [63].

Next, the investigation focuses on the reciprocal conversion of oxidized arsenic, with particular emphasis on the As_2O_3 molecule as the reference point. The three equations, specifically Eq. (6.10), (6.11), and (6.12), are under consideration.

$$2AsO + 1/2O_2 \rightarrow As_2O_3 \tag{10}$$

$$2AsO_2 \rightarrow As_2O_3 + 1/2O_2 \tag{11}$$

$$As_2O_3 + O_2 \rightarrow As_2O_5 \tag{12}$$

The reaction heat of the conversion of AsO to all three forms of As_2O_3, as described in (Eq. (10)), exhibits a negative value throughout the entire range of temperatures. As the temperature increases, the heat of reaction also increases, albeit remaining below -58 kcal mol^{-1}. This suggests that the presence of AsO in flue gas has a spontaneous tendency to produce As_2O_3. Furthermore, it has been observed that AsO undergoes a conversion to As_2O_3-1 at temperatures below 900 K, while at temperatures above 900 K, it transforms into As_2O_4. This finding aligns with the previously presented analysis. The heat of reaction for (Eq. (11)) demonstrates a consistently negative value across the entire range of temperatures. This observation suggests that the conversion of AsO_2 to As_2O_3 occurs spontaneously from a thermodynamic perspective. In comparison, it is noteworthy that the reaction heat of (Eq. (11)) is considerably higher than that of Equation (Eq. (6.10)), suggesting that the presence of AsO in flue gas poses a greater difficulty compared to AsO_2. Nevertheless, there are specific principles that dictate the conversion of As_2O_3 to As_2O_5. The thermodynamic spontaneity of the reaction described in (Eq. (12)) is limited to temperatures below 275 Kelvin. At temperatures exceeding 275 degrees Kelvin, the thermodynamic spontaneity of reaction (Eq. (12)) diminishes, resulting in the inhibition of the As_2O_3 to As_2O_5 conversion within the operational conditions of power facilities.

6. Removal of arsenic from high-temperature flue gas by photocatalysis

The photocatalytic reduction happens when the incident light energy (hv) is not less than the band gap (E_g), and the electron (e^-) in the valence band absorbs light energy and transitions to the conduction band, while the hole (h^+) is generated in the valence band. The generated electron (e^-) and the hole (h^+) migrate to the semiconductor surface respectively under the action of electric field or diffusion. The electron (e^-) with reducing ability and the hole (h^+) with oxidizing ability have redox reactions with the substances adsorbed on the semiconductor surface such as pollutant degradation, water decomposition to produce hydrogen, etc. [64]. Thus, it can be known that when the band gap is smaller, the electrons in the valence band are more easily to transition to the conduction band, and the photocatalytic reduction reaction is more likely to occur.

6.1 Mechanism of photocatalytic oxidation reaction

The electronic characteristics of semiconductors are reflected through their valence band (VB) and conduction band (CB). Semiconductor VB is by the highest occupied molecular orbital (HOMO) the interaction of form, and CB is a minimum of molecular orbital (LUMO) interact with each other. There are no electronic states between the top of VB and the bottom of CB. The energy range between CB and VB is called forbidden bandgap (also called energy gap or bandgap) and is often expressed as E_g. The band structure, including the band and the positions of VB and CB, determines the light absorption properties and redox ability of semiconductors and is one of the important properties of semiconductor photocatalysts [65].

After the photocatalyst is irradiated by ultraviolet and/or visible light (Vis) from sunlight or illumination sources, the electrons in the valence band are excited to the conduction band, while the holes remain in the valence band. Thus, this creates negative electron (e^-) and positive hole (h^+) pairs [66]. T these photoinduced electrons (or holes) first need to resist the recombination induced by hole capture (or electron capture), and then migrate through the body and surface of the photocatalyst to reach the catalytic reaction site to catalyze water decomposition, CO_2 reduction, and pollutant degradation [67]. n the photocatalysis process, e^-/h^+ pairs form in a few femtoseconds, their process from the original site to the reaction site takes only hundreds of picoseconds, and the catalytic reaction between e^-/h^+ and the adsorption reactants occurs in the time range of a few nanoseconds to a few microseconds [68]. The electron and hole recombination can last from a few picoseconds to tens of nanoseconds [69].

6.2 Photocatalytic removal of gaseous arsenic

Coal-fired power plants are the largest source of human arsenic pollution [54, 55, 70, 71], and effective control of various forms of arsenic emissions from coal-fired flue gas is critical to global arsenic pollution control. Arsenic in flue gas usually contains elemental arsenic and arsenic oxide such as AsO, AsO_2, As_2O_3, and As_2O_5 [54–57]. The As_2O_3-containing arsenic trivalent (As^{3+}) has stable thermodynamic properties, is difficult to dissociate, is highly toxic, and is the most harmful to the environment and human health [56, 72]. In addition, gaseous arsenic can interfere with selective catalytic reduction (SCR) devices and lead to arsenic poisoning of SCR catalysts, thereby reducing the economic efficiency of power plants [73, 74].

As^{3+} exists in the flue gas in the form of compounds, among which the simplest form is As_2O_3. Based on DFT calculation, As_2O_3 with different initial configurations was geometrically optimized, and the final configurations were shown in **Figure 6**.

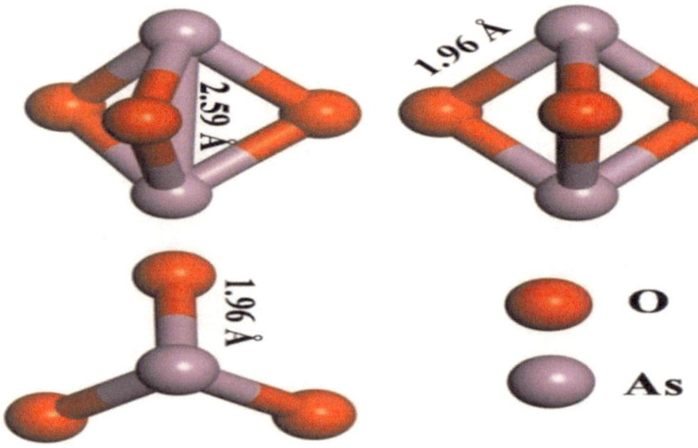

Figure 6.
Geometric optimization of gaseous arsenic As^{3+}.

Figure 7.
Schematic diagram of photocatalytic oxidation of gaseous arsenic As^{3+}.

As_2O_3, which is stable in the ground state, presents a spatially symmetric hexahedron structure. Each arsenic atom forms a single bond with three adjacent oxygen atoms and the As-O bond length is 1.96 Å. The strong bonding between arsenic and oxygen and the stability of molecular structure is not conducive to the transition of As^{3+} to higher valence state. The photocatalytic oxidation method is used to catalyze the oxidation of As^{3+} to a higher valence state. The possible reaction path is shown in **Figure** 7. The gaseous As^{3+} compound in the mainstream direction of the flue gas collides with the photocatalyst and has a certain probability of adsorption on the sample surface. The photocatalyst generates active groups such as hole (h^+), hydroxyl radical •OH, and superoxide radical $\bullet O_2^-$, which react with As^{3+} to oxidize it. Wu et al. developed Cu^{2+}-doped $BiOIO_3$ complex photocatalysts, rod-shaped Bi_2S_3 single-crystal adsorbents, In_2S_3/g-C_3N_4, CoS/g-C_3N_4, $Bi_4O_5I_2$/g-C_3N_5, and other series of complex photocatalysts/adsorbents around the effects of graphite-phase carbon nitride loading on the specific surface area and active sites of photocatalysts, as well as on the formation of bismuth-based heterojunctions, the effects of different sulfur doping on photocatalysts, and the adsorption characteristics of mercury and arsenic by photocatalysts under light-free or light conditions. CoS/g-C_3N_4, $Bi_4O_5I_2$/g-C_3N_5, and other series of complex photocatalysts/adsorbents. These adsorbents are characterized by increased specific surface area and smaller forbidden bandwidth, and it is pointed out that the adsorption characteristics of the photocatalysts for mercury, and arsenic are better in the presence of light conditions. The photocatalytic removal of arsenic from coal-fired flue gas is paved for the realization of photocatalytic removal.

7. Conclusion and outlook

Coal combustion in coal-fired power plants is one of the major anthropogenic sources of arsenic contamination in the environment. It has received increasing attention because of its high toxicity, volatility, and potential carcinogenicity. This chapter describes the nature and transport properties of arsenic in coal, as well as various arsenic removal techniques before, during, and after combustion. The main conclusions and outlook are as follows:

1. Arsenic can be divided into organic and inorganic arsenic, and there are two kinds of trivalent arsenic (As^{3+}) and pentavalent arsenic (As^{5+}), and the toxicity of As^{3+} is about 50 times that of As^{5+}. Arsenic in coal mainly exists in three forms: pyrite, organic arsenic, and arsenate. In the process of coal devolatilization, exchangeable and organically bound arsenic evaporates easily, compared with arsenate, which is very stable and usually decomposes at relatively high temperatures.

2. Arsenic is mainly bound to fly ash in the flue gas during migration. In the high-temperature flue, as the temperature decreases, the silicate melts in the fly ash may dissolve the arsenic and encapsulate it in the fly ash. In addition, gaseous arsenic in the flue gas interacts with metallic elements such as calcium, aluminum, and iron in the fly ash to form arsenates. The arsenic enriched in fly ash migrates with the flue gas to the low-temperature flue and is removed by wet FGD units, ESPs and wet ESP units, and fixed in the waste such as FGD gypsum, fly ash, and wastewater. Although air pollution control equipment can remove the vast majority of arsenic, there is still some arsenic escaping into the atmosphere in gaseous form or attached to submicron particles, as well as the removed gaseous arsenic entering the FGD wastewater mainly in the form of PM as arsenic in the liquid phase, which is more complex and difficult to remove. Therefore, improving the dust removal equipment or developing a greener technology will be a direction for future research.

3. There are many different control technologies available. These technologies cover methods for removing arsenic from flue gas before to, during, and after burning. Precombustion control technologies mainly use coal beneficiation technology to make the arsenic content in raw coal lower. However, this technology has limitations and is only applicable to control arsenic escape in the form of inorganic minerals. The in-combustion control technology uses the addition of arsenic fixing agents or mixed coal combustion to convert arsenic from gaseous to granular form and from fine to coarse particles for subsequent removal treatment, thus improving the effectiveness of arsenic removal. Post-combustion control technologies reduce arsenic emissions by using existing air pollution control equipment and sorbents to fix and remove arsenic from the flue gas. One of the more mature flue gas arsenic removal processes today is sorbent-based arsenic removal. The adsorbent is used to convert As_2O_3 to arsenic-containing particulate matter or to oxidize low-soluble and highly toxic As_2O_3 to highly soluble and less toxic As_2O_5, which is then captured by air pollution control equipment.

4. In order to remove arsenic from flue gas more effectively, many researchers have modified the adsorbents to obtain better applicability for arsenic removal in complex flue gas environments. It has been shown that the adsorption of gas-phase arsenic on CaO and Fe_2O_3 is mainly chemisorbed at 600–900°C. The adsorption amount and efficiency decreased with increasing temperature. The three most effective adsorbents for arsenic are iron oxide, calcium oxide, and aluminum oxide. To solve the problem of adsorbent sintering at high temperatures, adsorbents, such as CeO_2/CaO, and Al_2O_3/CaO, which are resistant to sintering, have been developed. At present, modification of adsorbents is a more effective way to improve the arsenic removal effect, and it can be an important direction to develop efficient arsenic adsorbents in future research.

5. The commonly used adsorbents for arsenic removal from coal-fired flue gas are relatively mature but have not yet been used in a wider range of applications and have high-cost problems. Therefore, there is an urgent need to develop new materials for the removal of arsenic. Photocatalysts have the advantages of mild reaction conditions, no secondary pollution, and strong redox properties. The application in arsenic removal from flue gas will be a great breakthrough.

Acknowledgements

This work was partially sponsored by the Nation Natural Science Foundation of China (52076126).

Funding

Partially Sponsored by National Natural Science Foundation of China (52076126)

Conflict of interest

The authors declare that they have no known competing financial interests or personal relationships that could have appeared to influence the work reported in this paper.

Author details

Jiang Wu*, Zhenzhen Guan and Yang Ling
College of Energy and Mechanical Engineering, Shanghai University of Electric Power, Shanghai, China

*Address all correspondence to: wjcfd2002@163.com

References

[1] Xing H, Liu H, Zhang X, et al. In-furnace control of arsenic vapor emissions using kaolinite during low-rank coal combustion: Influence of gaseous sodium compounds. Environmental Science & Technology. 2019;**53**(20):12113-12120

[2] Guan Y, Liu Y, Lv Q, et al. Bismuth-based photocatalyst for photocatalytic oxidation of flue gas mercury removal: A review. Journal of Hazardous Materials. 2021;**418**:126280

[3] Jia T, Wu J, Song J, et al. In situ self-growing 3D hierarchical BiOBr/$BiOIO_3$ Z-scheme heterojunction with rich oxygen vacancies and iodine ions as carriers transfer dual-channels for enhanced photocatalytic activity. Chemical Engineering Journal. 2020;**396**: 125258

[4] Liu X, Gao Z, Wang C, et al. Hg0 oxidation and SO_3, Pb0, PbO, $PbCl_2$ and As_2O_3 adsorption by graphene-based bimetallic catalyst ((Fe, Co)@ N-GN): A DFT study. Applied Surface Science. 2019;**496**:143686

[5] Bunt JR, Waanders FB. Trace element behaviour in the Sasol–Lurgi MK IV FBDB gasifier. Part 1–the volatile elements: Hg, As, Se, Cd and Pb. Fuel. 2008;**87**(12):2374-2387

[6] Liu H, Wang C, Zou C, et al. Simultaneous volatilization characteristics of arsenic and sulfur during isothermal coal combustion. Fuel. 2017;**203**:152-161

[7] Yu S, Zhang C, Ma L, et al. Deep insight into the effect of NaCl/HCl/SO_2/CO_2 in simulated flue gas on gas-phase arsenic adsorption over mineral oxide sorbents. Journal of Hazardous Materials. 2021;**403**:123617

[8] Kang Y, Liu G, Chou CL, et al. Arsenic in Chinese coals: Distribution, modes of occurrence, and environmental effects. Science of the Total Environment. 2011;**412**:1-13

[9] Tian HZ, Lu L, Hao JM, Gao JJ, Cheng K, Liu KY, et al. A review of key hazardous trace elements in Chinese coals: Abundance, occurrence, behavior during coal combustion and their environmental impacts. Energy & Fuels. 2013;**27**(2):601-614

[10] Yudovich YE, Ketris MP. Arsenic in coal: A review. International Journal of Coal Geology. 2005;**61**(3-4):141-196

[11] Goodarzi F. Inorganic constituents of coal and their impact on coal quality. CIM Bulletin. 1994;**87**(983):47-56

[12] Guo R, Yang J, Liu Z. Thermal and chemical stabilities of arsenic in three Chinese coals. Fuel Processing Technology. 2004;**85**(8-10):903-912

[13] Li CX. Environmental protection measures for coal-fired power plants. Clean the World. 2019;**10**:38-39 (in Chinese)

[14] Izquierdo M, Querol X. Leaching behaviour of elements from coal combustion fly ash: An overview. International Journal of Coal Geology. 2012;**94**:54-66

[15] Wang W, Qin Y, Song D, et al. Column leaching of coal and its combustion residues, Shizuishan, China. International Journal of Coal Geology. 2008;**75**(2):81-87

[16] Baba A, Gurdal G, Sengunalp F. Leaching characteristics of fly ash from fluidized bed combustion thermal power plant: Case study: Çan (Çanakkale-

Turkey). Fuel Processing Technology. 2010;**91**(9):1073-1080

[17] Otero-Rey JR, Mato-Fernández MJ, Moreda-Piñeiro J, et al. Influence of several experimental parameters on As and Se leaching from coal fly ash samples. Analytica Chimica Acta. 2005; **531**(2):299-305

[18] Senior CL, Helble JJ, Sarofim AF. Emissions of mercury, trace elements, and fine particles from stationary combustion sources. Fuel Processing Technology. 2000;**65**:263-288

[19] Sun JM, Yao Q, Liu HY. Distribution and enrichment mechanism of arsenic in inhalable particulate matter from coal burning. Journal of Coal science. 2004; **29**(1):78-82 (in Chinese)

[20] Contreras ML, Arostegui JM, Armesto L. Arsenic interactions during co-combustion processes based on thermodynamic equilibrium calculations. Fuel. 2009;**88**(3):539-546

[21] Chen D, Hu H, Xu Z, et al. Findings of proper temperatures for arsenic capture by CaO in the simulated flue gas with and without SO_2. Chemical Engineering Journal. 2015;**267**:201-206

[22] Wang Y, Yu J, Wang Z, et al. A review on arsenic removal from coal combustion: Advances, challenges and opportunities. Chemical Engineering Journal. 2021;**414**:128785

[23] Song DY, Qing Y, Zhang JY, Wang WF. Washability characteristics of hazardous trace elements in coals from western region of China. Journal of China University of Mining and Technology. 2006;**35**:255-282

[24] Guo X, Zheng CG, Liu YH. Study on the speciation of mercury, arsenic and selenium in coal. Fuel. 2001;**22**:763-766

[25] Akers D, Dospoy R. Role of coal cleaning in control of air toxics. Fuel Processing Technology. 1994;**39**(1-3): 73-86

[26] Wang WF, Qin Y, Song DY. Cleaning of potential hazardous elements during coal washing. Journal of Fuel Chemistry and Technology. 2003; **31**:295-299

[27] Zhou YP. Distribution type and occurrence form of arsenic in anthracite of laochang mining area. Coal Geology & Exploration. 1998;**26**:8-13

[28] Wang M, Song D, Zheng B, et al. The studying of washing of arsenic and sulfur from coals having different ranges of arsenic contents. Annals of the New York Academy of Sciences. 2008;**1140** (1):321-324

[29] Zhang Z, Fan J, Jin J, et al. Mode of occurrence of Pb, As, Be, Cr in coal. Ranliao Huaxue Xuebao (Journal of Fuel Chemistry and Technology);(China). 1992;**20**(2):206-212

[30] Gullett BK, Raghunathan K. Reduction of coal-based metal emissions by furnace sorbent injection. Energy & Fuels. 1994;**8**(5):1068-1076

[31] Zhao Y, Zhang J, Huang W, et al. Arsenic emission during combustion of high arsenic coals from southwestern Guizhou, China. Energy Conversion and Management. 2008;**49**(4): 615-624

[32] Zhao B, Chen G, Qin L, et al. Effect of coal blending on arsenic and fine particles emission during coal combustion. Journal of Cleaner Production. 2021;**311**:127645

[33] Liu Huimin，Wang Chunbo，Zhang Yue，et al. Experimental and modeling study on the volatilization of arsenic

during co-combustion of high arsenic lignite blends.Applied Thermal Engineering, 2016, 108: 1336-1343.

[34] Marczak M, Wierońska F, Burmistrz P, et al. Investigation of subbituminous coal and lignite combustion processes in terms of mercury and arsenic removal. Fuel. 2019;**251**:572-579

[35] Wu D, Liu J, Yang Y, et al. Experimental and theoretical study of arsenic removal by porous carbon from MSW incineration flue gas. Fuel. 2022; **312**:123000

[36] Bartoňová L, Čech B, Ruppenthalová L, et al. Effect of unburned carbon content in fly ash on the retention of 12 elements out of coal-combustion flue gas. Journal of Environmental Sciences. 2012;**24**(9):1624-1629

[37] Díaz-Somoano M, Martínez-Tarazona MR. Retention of trace elements using fly ash in a coal gasification flue gas. Journal of Chemical Technology & Biotechnology: International Research in Process, Environmental & Clean Technology. 2002;**77**(3):396-402

[38] Li S, Gong H, Hu H, et al. Re-using of coal-fired fly ash for arsenic vapors in-situ retention before SCR catalyst: Experiments and mechanisms. Chemosphere. 2020;**254**:126700

[39] Yu S, Zhang C, Ma L, et al. Insight into As_2O_3 adsorption characteristics by mineral oxide sorbents: Experimental and DFT study. Chemical Engineering Journal. 2021;**420**:127593

[40] Córdoba P, Ochoa-Gonzalez R, Font O, et al. Partitioning of trace inorganic elements in a coal-fired power plant equipped with a wet flue gas desulphurisation system. Fuel. 2012;**92**(1): 145-157

[41] Wang J, Zhang Y, Liu Z, et al. Effect of coordinated air pollution control devices in coal-fired power plants on arsenic emissions. Energy & Fuels. 2017; **31**(7):7309-7316

[42] López-Antón MA, Díaz-Somoano M, Spears DA, et al. Arsenic and selenium capture by fly ashes at low temperature. Environmental Science & Technology. 2006;**40**(12):3947-3951

[43] Diaz-Somoano M, Unterberger S, Hein KRG. Prediction of trace element volatility during co-combustion processes. Fuel. 2006;**85**(7-8):1087-1093

[44] Wang J, Zhang Y, Wang T, et al. Effect of modified fly ash injection on As, Se, and Pb emissions in coal-fired power plant. Chemical Engineering Journal. 2020;**380**:122561

[45] Zhang Y, Wang C, Li W, et al. Removal of gas-phase As_2O_3 by metal oxide adsorbents: Effects of experimental conditions and evaluation of adsorption mechanism. Energy & Fuels. 2015;**29**(10):6578-6585

[46] Hong J, Zhao Y, Wu J, et al. Fabrication of Al_2O_3/CaO with anti-sintering for efficient removal of As_2O_3 in simulated flue gas: Experimental and DFT study. Fuel. 2022;**307**:121812

[47] Zhao S, Xie X, Liao Y, et al. Removal of As_2O_3 in coal-fired flue gas by metal oxides: Effects of adsorption temperature and flue gas components. Journal of Cleaner Production. 2022;**376**: 134239

[48] Li S, Zhang Z, Liu Q, et al. Preparation of CeO_2/CaO with anti-sintering for efficient capture of As_2O_3 from flue gas at a high temperature. Energy & Fuels. 2021;**35**(24):20197-20205

[49] Zhang Y, Wang C, Liu H. Experiment and mechanism research on gas-phase As_2O_3 adsorption of Fe_2O_3/γ-Al_2O_3. Fuel. 2016;**181**:1034-1040

[50] Zhang Y, Liu J. Density functional theory study of arsenic adsorption on the Fe_2O_3 (001) surface. Energy & Fuels. 2019;**33**(2):1414-1421

[51] Zhao P, Guan J, Liu Q, et al. Microscopic spherical α-Fe_2O_3 for highly efficient gaseous arsenic capture in simulated flue gas under a wide temperature range. Energy & Fuels. 2021;**35**(23):19581-19591

[52] Ding W, Wan X, Zheng H, et al. Sulfite-assisted oxidation/adsorption coupled with a TiO_2 supported CuO composite for rapid arsenic removal: Performance and mechanistic studies. Journal of Hazardous Materials. 2021; **413**:125449

[53] Ling Y, Li J, Zou C, et al. Interaction mechanism between gaseous arsenic and the unburned carbon in coal-fired fly ash: A DFT combined thermodynamics study. Chemical Engineering Journal. 2021;**425**:130714

[54] He KQ, Yuan CG, Jiang YH, et al. Synergistic effects of Fe-Mn binary oxide for gaseous arsenic removal in flue gas. Ecotoxicology and Environmental Safety. 2021;**207**:111491

[55] Tian C, Gupta R, Zhao Y, et al. Release behaviors of arsenic in fine particles generated from a typical high-arsenic coal at a high temperature. Energy & Fuels. 2016;**30**(8):6201-6209

[56] Shen F, Liu J, Zhang Z, et al. On-line analysis and kinetic behavior of arsenic release during coal combustion and pyrolysis. Environmental Science & Technology. 2015;**49**(22): 13716-13723

[57] Winter RM, Mallepalli RR, Hellem KP, et al. Determination of As, Cd, Cr, and Pb species formed in a combustion environment. Combustion Science and Technology. 1994;**101**(1-6):45-58

[58] Perdew JP, Burke K, Ernzerhof M. Generalized gradient approximation made simple. Physical Review Letters. 1996;**77**(18):3865

[59] Ernzerhof M, Perdew JP. Generalized gradient approximation to the angle-and system-averaged exchange hole. The Journal of Chemical Physics. 1998;**109**(9):3313-3320

[60] Xu W, Zhang C, Shen H, et al. Surface oxalate coordination: Facilitating basic OER activity by regulating the electronic structure of Fe–Ni_3S_2. ACS Sustainable Chemistry & Engineering. 2022;**44**(10):14396-14406

[61] Zou C, Wang C, Chen L, et al. The effect of H_2O on formation mechanism of arsenic oxide during arsenopyrite oxidation: Experimental and theoretical analysis. Chemical Engineering Journal. 2020;**392**:123648

[62] Ling C, Shi L, Ouyang Y, et al. Nanosheet supported single-metal atom bifunctional catalyst for overall water splitting. Nano Letters. 2017;**17**(8):5133-5139

[63] Gao Z, Zhao M, Yan G, et al. Identifying the active sites of carbonaceous surface for the adsorption of gaseous arsenic trioxide: A theoretical study. Chemical Engineering Journal. 2020;**402**:125800

[64] Chong MN, Jin B, Chow CWK, et al. Recent developments in photocatalytic water treatment technology: A review. Water Research. 2010;**44**(10): 2997-3027

[65] Ma Y, Wang X, Jia Y, et al. Titanium dioxide-based nanomaterials for photocatalytic fuel generations. Chemical Reviews. 2014;**114**(19):9987-10043

[66] Bai S, Jiang J, Zhang Q, et al. Steering charge kinetics in photocatalysis: Intersection of materials syntheses, characterization techniques and theoretical simulations. Chemical Society Reviews. 2015;**44**(10):2893-2939

[67] Gopidas KR, Kamat PV. Photoinduced charge transfer processes in ultrasmall semiconductor clusters. Photophysical properties of CdS clusters in Nafion membrane. Journal of Chemical Sciences. 1993;**105**:505-512

[68] Zhang JZ. Interfacial charge carrier dynamics of colloidal semiconductor nanoparticles. The Journal of Physical Chemistry B. 2000;**104**(31):7239-7253

[69] Kubacka A, Fernandez-Garcia M, Colon G. Advanced nanoarchitectures for solar photocatalytic applications. Chemical Reviews. 2012;**112**(3):1555-1614

[70] Wang C, Liu H, Zhang Y, et al. Review of arsenic behavior during coal combustion: Volatilization, transformation, emission and removal technologies. Progress in Energy and Combustion Science. 2018;**68**:1-28

[71] Yang W, Gao Z, Liu X, et al. The adsorption characteristics of As_2O_3, Pb^0, PbO and $PbCl_2$ on single atom iron adsorbent with graphene-based substrates. Chemical Engineering Journal. 2019;**361**:304-313

[72] Liu H, Wang C, Zou C, et al. Vaporization model of arsenic during single-particle coal combustion: Numerical simulation. Fuel. 2021;**287**: 119412

[73] Zhang H, Kong M, Cai Z, et al. Synergistic effect of arsenic and different potassium species on V_2O_5-WO_3/TiO_2 catalyst poisoning: Comparison of Cl^-, SO_4^{2-} and NO_3^- anions. Catalysis Communications. 2020;**144**:106069

[74] Lu Q, Pei X, Wu Y, et al. Deactivation mechanism of the commercial V_2O_5–MoO_3/TiO_2 selective catalytic reduction catalyst by arsenic poisoning in coal-fired power plants. Energy & Fuels. 2020;**34**(4):4865-4873

Section 2

Impacts

Chapter 3

Consequences of Arsenic Exposure in Plant-Health Status: An Overview

Sweety Nath Barbhuiya, Ringthardi Warisa, Dharmeswar Barhoi, Jurimoni Das and Sarbani Giri

Abstract

Arsenic is the biggest threat to all living organisms across the world. It is typically present in very minute amounts in rock, soil, air, and water, but these levels are rising as a result of both natural and man-made activity. Exposure to arsenic increases the risk of developing liver, lung, kidney, and bladder malignancies as well as vascular illnesses such as stroke, ischemic heart disease, and peripheral vascular disease. Arsenic generates oxidative stress, which disrupts the redox balance. In fact, in plants arsenic gets accumulated in different parts of plants upon exposure to either contaminated soil or water, causing hazardous effects on the plant. Therefore, this chapter is aimed to understand the effect of arsenic exposure on the growth and development of the plant as a whole.

Keywords: arsenic, plant health, human health, toxicity, arsenic metabolism

1. Introduction

Arsenic is a ubiquitous metalloid found all across the earth's crust and is considered to be highly toxic to all living organisms [1, 2]. The arsenic concentration of 10 μg/l in drinking water is considered to be safe as per the recommendation of the World Health Organization and the U.S. Environmental Protection Agency (EPA) [3]. But many countries like China, Chile, Argentina, Vietnam, Canada, Laos, Mexico, Ghana, many parts of the United States, Bangladesh, and India have high concentrations of arsenic in their groundwater [3–7]. This elevated concentration of arsenic in the groundwater level is caused by both anthropogenic activities as well as natural sources. The International Agency of Research on Cancer (IARC) classified arsenic as a class I human carcinogen and also ranks 1st on the US Agency for Toxic Substances and Disease Registry (ATSDR) Priority List of Hazardous Substances (https://www.atsdr.cdc.gov/spl/index.html) [8]. Arsenic exposure can cause various hazardous effects in human beings causing diseases like diabetes, hypertension, cardiovascular disease, gastrointestinal, renal, and neurological disorders, skin keratosis, and cancer [6, 9, 10]. In Bangladesh, where arsenic contamination at the groundwater level is extremely elevated, incidences of "black foot disease" is reported to be very high [11]. Arsenic even enters the food chain of human beings through the food chain via contaminated crops. There are different ways by which arsenic gets into a farming system that includes natural geochemical processes, irrigation using arsenic-contaminated water, and the use of pesticides containing arsenic [12, 13]. Arsenic can induce toxicity in plants as characterized by decreased germination; decreased plant biomass and chlorophyll content [14].

Thus, this chapter is aimed to understand the mechanism by which arsenic gets accumulated in the plants and thereby causing toxicity at the cellular level, and eventually degrading the well-being of the exposed plants.

2. Types of arsenic

In the water and soil ecosystems, arsenic can appear in three primary forms: inorganic, organic, and arsine gas (−3 oxidative state). It also has three main valence states: arsenic element (0), arsenite (trivalent +3), and arsenate (pentavalent +5) [1]. Both organic (monomethyl arsenic) and inorganic (arsenite) trivalent arsenic compounds are often viewed as being more hazardous than pentavalent ones. Compared to As V, As III is 60 times more poisonous [10]. The chemical compounds arsenate As(V), arsenite As(III), monomethyl arsenic acid (MMAC), dimethyl arsenic acid (DDMA), trimethylarsine (TMA), arsenocholine (AsC), arsenobetaine (AsB), and arsenosugars can all be found in the environment (**Figure 1**) [15]. Environmental versions include methyl arsenic acid, dimethyl arsenic acid, arsine, etc., and also consist of arsenious acids (H_3AsO_3, $H_3AsO_3^{2-}$), arsenic acids (H_3AsO_4, $H3AsO_4^{-}$, $H_3AsO_4^{2-}$),

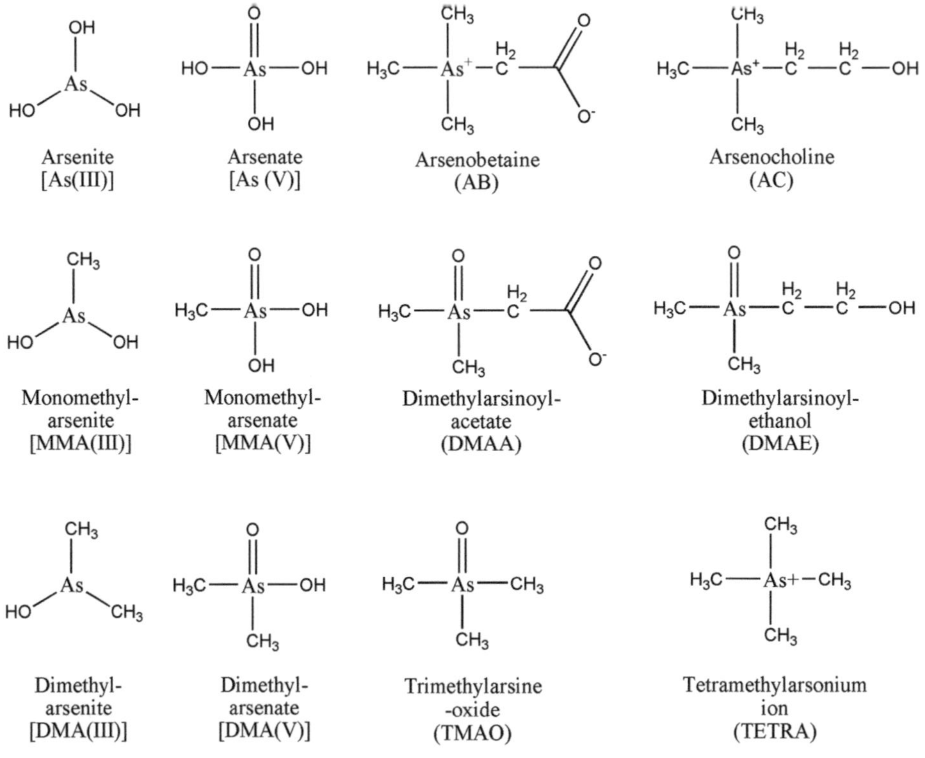

Figure 1.
Chemical structure of As species detected in terrestrial plants [15].

arsenates and arsenates [16]. The Marine invertebrates such as lobsters, brown algae, seaweed, fish, and seawater bivalves are the principal sources of the organic arsenic species (MMA, DMA, AsC, and AsB). As(V) and As(III) are the predominant arsenic species in inorganic terrestrial settings, however small amounts of organic arsenic species have been discovered as a result of microbial activity or the application of organic herbicides and insecticides [15]. Arsenate As(V) dominates in the soil solution under oxidizing conditions, whereas As(III) dominates under mildly reducing conditions. In contrast to organic arsenic, inorganic arsenic is poisonous. Most people believe that inorganic arsenic is more hazardous than organic arsenic [17].

Arsine gas is the form of arsenic that is the most dangerous. While concentrations over 25 ppm are reported to be fatal in under an hour of exposure and concentrations over 250 ppm are stated to be instantly fatal, concentrations above 10 ppm are said to be fatal upon inhalation. Arsine gas is colorless, does not irritate tissues, and has a hardly perceptible odor. In terrestrial ecosystems, arsenic can be found in four oxidation states (III, 0, +III, and +V). Under aerobic soil conditions and anaerobic circumstances (submerged soils), respectively, arsenate (AsV) and arsenite (AsIII) are the prevalent forms [16]. The interconversion of these two arsenic forms (arsenate and arsenite) is highly influenced by both biotic (microorganisms) and abiotic (changes in redox potential and pH) variables [18, 19].

3. Sources of arsenic exposure

3.1 Natural sources

3.1.1 Geologic sources

The geological features of the area are frequently linked to the presence of arsenic in groundwater. For example, in Bangladesh and West Bengal, the source of arsenic in groundwater is linked to the presence of arsenic-bearing minerals in the Ganges delta deposits [20].

3.1.2 Contaminated soil

Natural processes like the weathering of rocks and minerals can cause soil to become contaminated with arsenic. But in some places, human endeavors like mining have markedly raised the amount of arsenic in the soil [21].

3.1.3 Drinking water

Due to the local geological features, arsenic can be found in natural water sources including rivers and groundwater. For instance, arsenic in groundwater is a substantial public health concern in many regions of South Asia [22, 23].

3.2 Anthropogenic sources

3.2.1 Industrial activities

Industrial processes including mining, smelting, and the use of pesticides are examples of anthropogenic sources of arsenic. For instance, major arsenic poisoning of soil and water has been related to the use of arsenic in pesticides [24].

3.2.2 Food

Due to the usage of arsenic-containing pesticides or polluted soil, arsenic may be found in food. Due to irrigation with arsenic-contaminated water, it has been discovered that rice in some locations contains high levels of arsenic [25].

3.2.3 Wood preservatives

Up until the early 2000s, arsenic was often utilized in wood preservatives, resulting in significant arsenic pollution of the soil and water around treated wood sites [26].

4. Arsenic-induced toxicity in human health

Clinical symptom development differs depending on whether arsenic exposure was acute or persistent. However, clinical signs of acute exposure to arsenic emerge considerably more quickly than those of chronic exposure, which take a longer time to manifest. Organ damage and possible death are possible in acute arsenic intoxication. On the other hand, malignant tumors may develop as a result of the deformity of the extremities brought on by long-term exposure to arsenic [23, 27]. Ingestion and inhalation are the two most common ways to be exposed to arsenic, and these are also the ways that cause health issues most commonly. Dermal exposure can also result in sickness, but it is less likely to do so than ingesting or inhaling arsenic. Based on how arsenic affects different organs such as the skin, brain, heart, pancreas, lungs, liver, and kidney, arsenic-induced toxicity in humans can be studied. A detailed study of it is discussed below.

4.1 Skin

Skin conditions are the early signs of arsenic intoxication. The consequences of arsenic exposure on health have been the subject of numerous researches. For instance, a study carried out in Bangladesh revealed that drinking water contaminated with arsenic over time was linked to an increased risk of skin lesions and skin cancer [2, 21, 28].

4.2 Heart

An elevated risk of cardiovascular disease was linked to long-term exposure to arsenic in drinking water, according to a systematic review and meta-analysis of epidemiological data [2, 6, 29]. An increase in the number of deaths from aneurysms, arteriosclerosis, and other related diseases was observed in another study carried out in the USA in locations where the drinking water contained high arsenic contents >20 g/l. [30]. Wang et al. [31] discovered that Taiwanese communities living in areas with arsenic-polluted wells (>0.35 mg/l) had an elevated incidence of illness in the blood vessels.

4.3 Pancreas

A study by Hassan et al. discovered a direct correlation between eating rice that has been exposed to arsenic contamination and developing diabetes. For the two billion people who live in Asia, rice is the main food source. According to studies, rice has high levels of inorganic arsenic, which can damage pancreatic beta cells and interfere with the body's ability to regulate glucose levels [32].

4.4 Brain and the nervous system

The central nervous system is the primary target organ for the harmful effects of heavy metals like arsenic [27]. A case–control study involving 57 people who had arsenic-induced skin lesions and had long-term exposure to arsenic-contaminated drinking water was conducted in China. According to the study, exposed individuals experienced a variety of neurological problems, including aberrant distal feeling, decreased temperature and pressure perception, and functional lesions of the vegetative nerves, such as hypohidrosis and adiaphoresis [33]. In a recent study conducted in West Bengal, India, patients exposed to drinking groundwater tainted with arsenic experienced peripheral neuropathy as their main neurological consequence [34].

4.5 Lungs

Both prenatal and childhood exposure to arsenic through drinking water was associated with both long-term lung function and non-malignant lung illnesses in humans, according to Dauphiné et al. [35]. Respiratory problems such as persistent cough, laryngitis, bronchitis, and rhinitis are frequently caused by inhaling arsenic dust or fumes while mining or milling ores [36]. Workers exposed predominantly to arsenic trioxide dust have been found to have an elevated incidence of lung cancer [37, 38].

4.6 Liver

Following repeated exposures, there is a proportionally higher accumulation of arsenic in the liver, which makes it more harmful to the liver [10]. Some of the early clinical signs of liver illness brought on by exposure to arsenic include esophageal varices bleeding, ascites, jaundice, or liver enlargement. Hepatic lesions may develop in the later stages of severe toxicity coupled with other problems such as cirrhosis with a high likelihood of liver failure, non-cirrhotic portal fibrosis, and hepatic fibrosis [39].

4.7 Kidney

Renal injury (acute tubular necrosis) caused by arsenic toxicity is one of the clinical symptoms, which is followed by hypogea, high serum creatinine levels, blood urea nitrogen, and proteinuria [40].

5. Arsenic exposure to animals and humans through the food chain

Millions of people are exposed to arsenic through food, air, water, and soil, leading to adverse long-term health consequences. It is important to monitor the food chain as humans are the top consumers, making it a challenging task. Arsenic enters the ecosystem through several processes, including geogenic phenomena (thermal regime, volcanic activities, weathering, and leaching), anthropogenic activities (urbanization, industrial establishments, groundwater extraction for irrigation, extensive use of pesticides and fertilizers), and biogenic (induced by plants, animals, microorganisms, and aquatic biota) phenomena. Due to its significant industrial importance, the use of As is unavoidable as it is employed in the production of semiconductors, explosives, munitions, paints, cosmetics, insecticides, herbicides, and fertilizers, among other things [41]. Food security is threatened and human health is further jeopardized by arsenic entering the food chain and affecting crop quality and production. Chronic exposure to arsenic can seriously deteriorate health across the food pyramid. In addition to overall As concentration, bioavailability also plays a significant role in the accumulation of arsenic from one trophic level to the next. Since humans may also absorb arsenic via tainted rice, vegetables, milk, and meat, "plant-human" and "plant–animal-human" could be additional possible food chain routes for arsenic buildup.

Arsenic may enter the body through a number of different routes, including drinking and cooking water, crops and vegetables grown in areas with high arsenic levels, and animal products (meat, milk, and eggs) [42]. It has been determined that two key pathways for its entry into the food chain are through the consumption of contaminated crops and drinking water. Due to its high water solubility and propensity for bioaccumulation in several environmental matrices, arsenic becomes exceedingly hazardous even at low exposure levels. Only the bioavailable form of arsenic enters the body directly after being consumed in tainted food or water, interfering with many metabolic processes. The process by which plants absorb arsenic changes according to chemical speciation. As(V) has reportedly been shown to enter plant cells through inorganic Phosphate (Pi) channels. However, because their chemical structures are similar, plants may absorb As(III) through a variety of intrinsic proteins that resemble nodulin-26 and silicon transporters. Whether in the root or shoot tissue, plants normally store the majority of arsenic in vacuolar compartments. Animals can absorb As from a variety of sources, but they mostly do so from the soil matrix. They are exposed to elevated levels of arsenic, and continuous consumption of such crops is associated with a number of health risks [43].

One of the worst environmental catastrophes in the previous century was chronic arsenic poisoning brought on by drinking water. In India, groundwater arsenic contamination was first reported in West Bengal in 1983. With a concentration of up to 4000 g/l, the Ganges-Brahmaputra-Meghna plain is the most severely arsenic-polluted place in the world. Besides drinking, rice cooked using contaminated groundwater contains more arsenic than uncooked rice due to the chelation of As by rice grain and evaporation of water while cooking, which concentrates the metalloid in the remaining water [42].

In addition to drinking water, vegetables, and other crops cultivated in groundwater contaminated with arsenic can significantly increase the daily consumption of arsenic in human meals. Agriculture is the primary industry in West Bengal's heavily arsenic-affected districts, therefore over the past few decades, unrestricted groundwater pumping for irrigation has been a common procedure. A substantial quantity of arsenic is accumulated in West Bengal by the crops and vegetables grown in soil contaminated with arsenic, and this arsenic eventually makes its way into the food chain people. The arsenic toxicity is disseminated through the fruits, vegetables, and grains that are grown on soil that contains arsenic. In some countries, it has been found that rice has a 10-fold greater As content than other cereal grains [43]. In comparison to wheat (0.1) and barley (0.2), rice has a higher arsenic translocation factor (TF) of 0.8. Elements that affect the amount of arsenic in rice are how the grain is processed, the kind of rice farm, the area where it is grown, the irrigation system used, and the way of cooking.

Food crops including cereals and vegetables have been considered key conduits for arsenic exposure in humans due to plants' intrinsic capacity to collect minerals from the soil [41]. Vegetables can be exposed to As due to either natural or anthropogenic activitiesIt is well known that irrigation water from polluted groundwater resources may introduce arsenic into the food chain. Vegetable kind, soil type, irrigation water quality, proximity region, human pressure, and geogenic contamination are some of the variables affecting arsenic contamination in vegetables [44]. Arsenic levels in vegetables were found to be above both national and international standards in research from the hamlet of Samta in West Bengal, India. Tube well water used for irrigation that had 0.24 mg/l of arsenic was linked to the contamination in plants. Compared to its abundance in roots and other below-ground plant organs, arsenic is found in relatively lower concentrations in grains, seeds, lentils, and fruits. Additionally, root vegetables' exterior root skin contains more arsenic than the actual root itself does.

Arsenic enters the food chain through arsenic-contaminated water and feed for cattle. Cattle feeding practices, food, water, and environmental contaminants all affect the amount of arsenic in the milk they produce High arsenic concentration in raw milk was reported from cows grazing near lava ground in Turkey. In Pakistan, elevated arsenic concentration was found in milk samples of different milch animals due to contaminated drinking water (geogenic arsenic). This is likely due to the contaminated drinking water given to cattle [41]. The content of As in the scalp hair of children of three age groups was high, with goat and sheep milk being significantly higher than cows and camel, suggesting that the main exposure of As is not only due to milk [45].

Commercial chicken meat is a common source of animal protein and is consumed globally. Despite being nutritious, the quality of poultry meat may be compromised by the presence of harmful metals caused by a variety of human activities. The use of organo-arsenic pharmaceuticals as animal feed and antiparasitic treatments encourages the transmission of arsenic in cattle and poultry products. Roxarsone is linked to elevated arsenic levels in chicken tissues [41]. High arsenic levels were reported in poultry meat in Bangladesh due to drinking contaminated groundwater [46].

Arsenic is also abundant in seafood at concentrations as high as several hundred micrograms per gram [47]. Organic arsenic, including arsenosugars, arsenobetaine, and arsenolipids, is mostly found in seafood. The most prevalent arsenic substances in fish and seaweed are arsenobetaine and arsenosugars. Arsenic poisoning is more prevalent among coastal populations whose diets are mostly composed of seafood. Arsenic poisoning is more common among coastal populations whose diets are mostly composed of seafood, with higher levels of arsenic in the blood and urine, baby cord blood, and breast milk [48].

6. Arsenic metabolism in plants

Most plants are dangerously affected by arsenic exposure at lower or even greater doses. Exposure to arsenic can interfere with metabolic processes and inhibit plant development [49]. When examining the effects of arsenic on plant cellular metabolism, it's important to take into account the various arsenic species that are present in soils, their capacity to enter plant cells, their potential to change from one arsenic species to another, and the different arsenic transport pathways that are present within the plant. Cell types exposed to high concentrations of specific arsenic species can be used to discriminate between hyperaccumulators and non-accumulators.

While non-accumulators prefer to reserve arsenic in root cells and have much lower quantities of arsenic in shoot cells, hyperaccumulators have abnormally high concentrations of arsenic in the cells of aerial tissues compared to the root [50]. The metabolism of arsenic has a significant impact on how dangerous it is [51]. According to research done by Martinez-Castillo et al. [52] As interacts with a variety of metabolic processes, causing physiological and morphological abnormalities that affect plant development (poor nutrient absorption) and productivity (reduced seed germination), while also inhibiting root growth and cell death.

The first step in the detoxification of arsenate is the reduction of arsenate to arsenite after arsenic has entered plant root cells [18, 53]. According to a report, several enzymes from various systems exhibit AsV reductase activity [18, 50]. They include glycogen phosphorylase, glyceraldehyde-3-phosphate dehydrogenase (GAPDH), polynucleotide phosphorylase, purine nucleoside phosphorylase (PNP), and mitochondrial F1Fo ATP synthase. As V is first reduced to As III to start As a metabolism, this reduction can happen both enzymatically and without the use of an enzyme. As V is converted to As III through a non-enzymatic process in which two molecules of reduced glutathione (GSH) are oxidized to produce oxidized GSH, which is then converted back to two GSH molecules with the aid of GSH reductase. But the reduction of arsenate is a very slow process. Arsenate reductase directly converts As V to As III in enzymatic processes. According to a study, arsenic primarily enters plants in an inorganic form, As(III) or As(V), through transporter proteins. This process is likely controlled by a gradient of arsenic concentration between plant cells and growth media [54]. AsV is quickly converted into AsIII, the more hazardous of the two forms, once it penetrates plant cells [55]. Once present in the soil, the two inorganic forms of arsenic, arsenate (AsV) and arsenite (AsIII) are promptly absorbed by the plant roots' cells. According to reports, plants are extremely capable of converting AsV and MMA(V) that their roots have ingested into AsIII and MMAIII, respectively. When MMA(III) was completely absent from the medium, it was found that MMAV was transformed into MMA(III) within the roots (**Figure 2**) [51]. The reduction of a trivalent state and oxidative methylation to a pentavalent state are the two main steps in the metabolism of arsenic in plants. The plant ACR2 gene, a homolog of the yeast arsenate reductase, was thought to cause As(V) reduction in plant cells. HAC1 (for

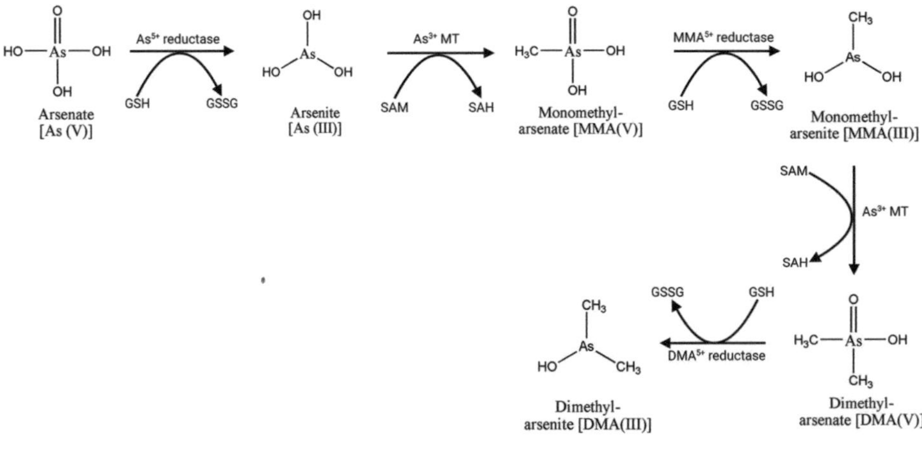

Figure 2.
Pathways for arsenic methylation in plant [51].

High As Content 1) or ATQ1 (for Arsenate Tolerance QTL1), a new As(V) reductase, has recently been found in Arabidopsis. The ability of the protein from the rhodanese-like family to reduce As(V) to As(III) HAC1/ATQ1 both in vitro and in planta has been described [56].

Compared to trivalent arsenicals, pentavalent arsenicals are less hazardous [17]. Any type of arsenic that is absorbed by plants causes the cellular conversion of inorganic arsenic to organic species. The enzyme arsenate reductase speeds up the process of conversion of AsV to AsIII in the plant cell after arsenic absorption. AsIII, on the other hand, destroys the functionality of proteins by creating a molecule that contains their thiol groups. Nevertheless, GSH in vitro facilitates this reduction non-enzymatically [52].

Arsenic-induced phytotoxicity has negative effects on plant growth, development, and metabolism, which lower plant output [1, 57]. Even though the absorption mechanisms for arsenite and arsenate are different, arsenite was successfully transported as arsenate across the plasmalemma [58].

Inorganic arsenic only occasionally travels through the majority of plants. From the roots to the stems to the leaves to the grain, there was a dramatic fall in the concentration of As. AsV is most likely to be carried by Pi transporters, although AsIII was found to be transferred across membranes by silicon transport proteins such as OsLsi1 and OsLsi2 [50].

7. Bioaccumulation of arsenic in plants

In soils, arsenic may be present in a wide range of chemical forms, such as exchangeable, easily soluble (slightly bound to carbonate and particle surface), poorly crystalline, amorphous, and crystalline Fe and Mn oxide, Fe sulfide, and organic-matter-bound phases, as well as incorporated in crystal structures [59]. The amount of arsenic that accumulates in plants is dependent on many factors the physicochemical characteristics of the soil, which have a detrimental effect on plant growth, and nutrients that restrict plant growth and the bioavailability of arsenic [60]. Several parts of plants exposed to arsenic-contaminated soil and water have been found to contain arsenic.

Arsenic can penetrate terrestrial and aquatic ecosystems naturally or as a result of anthropogenic activity. These ecosystems have a route in the roots that permit plants to take up arsenic along with nutrients. Depending upon the variety of plants (whether As hyperaccumulator/non-hyperaccumulator or arsenic tolerant/susceptible), the mechanisms of arsenic accumulation, absorption, or toxicity response may be different. A large amount of As flow occurs through the xylem or phloem and may be transported to different parts of plants including the stem, leaf, reproductive portions, and even seeds after inflowing the root epidermis and moving through the apoplastic and symplastic pathways. Arsenic accumulation in leaves and roots is directly proportional to the concentrations of AsV and AsIII, except at the peak of AsIII concentration, probably due to AsIII extrusion [61]. However, certain plants, known as hyperaccumulating plants, can metabolize high concentrations of arsenic without experiencing any physiological damage, while others can get damaged from even small doses of arsenic exposure.

Arsenic concentration in plants varies considerably depending on the type of plant, the nature of the soil, and the concentration of arsenic in the irrigation water [62]. It has been revealed that roots are more accountable in terms of arsenic accumulation. It was reported the main arsenic species in aerobic soils is arsenate [18]. The solubility of arsenic is highly affected by the soil properties Regardless of the overall As content of the soil, soil properties such as pH, organic carbon, texture, and minerals, apart from the arsenic content of the soil [60, 63]. Arsenic uptake by plants is dependent mainly on the source, pedology, species, and age. Even though the quantity of absorption by plants differs based on several conditions, it was evident that soil level is directly proportional to the concentrations of inorganic arsenic, both Arsenate As(V) and arsenite As(III), which highly prevail in plant tissue, however, a trace amount of organic arsenic species (<1%) has been found in the shoots. It was revealed that the vegetables with juicy content attain the maximum concentrations of arsenic, while the fruits with slight juicy content have a diminished concentration of arsenic [64].

Arsenic contamination is the most important problem worldwide associated with either groundwater or irrigation of rice cultivated in affected areas [9]. Rice is the most important crop for providing essential nourishment to 50% of the world's population. However, rice has the propensity to accumulate heavy metals to a greater extent, especially arsenic, thus, having a hazardous effect on human health [65, 66]. In comparison to other cereal crops that are used as staples, rice is faraway efficient in absorbing arsenic into the grains. The bioavailability of arsenic in soil is soaring as because rice is normally produced under submerged, flooded conditions. Since some arsenic species are phytotoxic, arsenic may hurt the volume of rice formed overall. The sequence of the toxicities of the various arsenic species is AsH_3 > As(III) > As(V) > MMAA (monomethyl arsenic acid) > DMAA (dimethyl arsenic acid) [62]. The main crucial form of arsenic species available in paddy fields is arsenite, but a small amount of arsenate, MMAA, and DMAA is also found [60]. The ingress of MMA(V) and DMA(V) in *O. sativa* has been suggested to engross aquaporin Lsi1. It is reported that the As(III) is taken up by the root more rapidly than other organic species of arsenic. It was suggested that Lsi1 is responsible for the entry of As(III) in the plant, while Lsi2 is responsible for the efflux of As(III) from the root to the xylem, as evident from the study in *O. sativa L.* The arsenic concentration present in the soil or growth media has a great impact on the uptake and accumulation of arsenic by plants and the levels of arsenic significantly rise with augmentation in levels of arsenic [67, 68]. In the year 1998, Xie and Huang [68] reported the order of arsenic concentration in different parts of rice plants and it was as follows: root > straw > husk > whole grain > husked rice. The maximum accumulation of arsenic was reported in the roots of the plants in comparison to the other parts [69].

In reducing conditions, arsenite As(III) is the principal As species, such as in flooded paddy soils. The bioavailability of arsenic in rice plants is elevated by moving the arsenite into the soil solution by flooding the paddy fields and arsenite may be rapidly absorbed by the roots of the plants from the surrounding environment. In the roots of rice as well as of other aquatic plants whose roots flourish in anaerobic or semi-anaerobic conditions, the absorption of arsenite is particularly very important [18]. According to a recent study, the glycerol-transporting channel facilitates the uptake of undissociated (III) (pKa 9.2) in *Saccharomyces cerevisiae* (baker's yeast) and also helps in the uptake of arsenite in *Oryza sativa* (rice). The uptake of As(III) and As(V) is therefore carried out by a range of plasma membrane mechanisms [14]. In general, As(III) and As(V) uptake is elevated than that of DMA and MMA in plant systems.

The Chinese brake (Pterisvittata L.) fern can accumulate up to 23 g of arsenic per kilogram owing to its growth on arsenic-contaminated soil [70], while the concentrations of As in other varieties of terrestrial plant shoots are tremendously low. This plant has immense capacity to transfer arsenic to its above-ground biomass (up to 90% of the total As intake), in addition to absorbing a considerable quantity of As from the soil (up to 2.3% of dry plant weight).

Comparatively speaking to the As chemical form prevalent in soil solution, rice (O. sativa, L.) absorbs arsenic. Most of the arsenic is present as As when the soil redox potential falls below 0 mV. (III). Both As(III) and As(V) are present in more oxidizing circumstances. The photo availability of As in the water-soluble fraction is influenced by its chemical speciation. The arsenic content [As(III)] in the plant is increased with rising arsenic concentrations in solutions and rising levels of soluble arsenic [67].

Several arsenic hyperaccumulators have been reported, including *Holcuslanatus* and *Vallisnerianeotropicalis*. Many macrophytes have also been engaged for arsenic bioaccumulation. A study indicated that in plants water relation is regulated by arsenic stress, diminishing relative water content, the efficiency of water use, stomatal conductance, and rate of transpiration. The content of leaf water may also get hampered as a result of disruption in the cell wall structures of the leaf. It was established that the high affinity for the absorption of phosphate can increase the absorption of arsenate [71]. Due to the same chemical characteristics that exist between As(V) and P(V), both of them may be transported via the identical plasma membrane transport mechanism. Moreover, it was hypothesized that aquaporins in the roots are involved in As(III) absorption. A fraction of the quantity of As(V) that readily enters the plant is reduced to arsenite, which leads to the generation of oxidative stress in plants.

8. Consequences of arsenic exposure in plants

Some harmful exposures come from naturally existing compounds, like the arsenic frequently present in drinking water, therefore environmental health problems are not just restricted to toxic waste sites and poisoning incidents. Arsenic contamination of soil can impair a plant's ability to function normally, resulting in stunted growth and subpar agricultural output [54]. Arsenic can still have an effect on several biological systems for years or even decades after exposure levels have diminished. Arsenic generates oxidative stress, which disrupts the redox system [72]. The world's two main sources of arsenic pollution are groundwater and the irrigation of rice farmed in polluted areas. When arsenic levels in irrigation water or soil are elevated, it may hinder normal plant growth by causing symptoms including lower biomass in root and shoot wilting and necrosis of leaf blades, decreased leaf area and photosynthesis, and decreased fruit and grain output [62]. Several artificial arsenic compounds were utilized in agriculture as efficient pests, parasites, and weed control tools, and they slowly accumulated in the soil. Moreover, certain additional morphological processes may cause the quantity of arsenic in the environment to grow, thus increasing human exposure to arsenic [73]. ROS may be produced as a result of arsenate toxicity, which then damages membranes. Lipid peroxidation may cause structural cell membrane damage and accelerate ionic exosmosis [17]. Arsenic frequently has the effect of shortening the root and shoot length. Because plant roots are the initial point of contact for these lethal arsenic species in the nutritional media, the growth of root length was noticeably hindered when the arsenic content rose. It is well known that modest concentrations of arsenicals can alter the phosphorylation states of signal transduction proteins, resulting in the start of gene transcription. Arsenic disrupts phosphorylation via activating specific phosphatases, thiol-dependent phosphatases, or phosphotransferase activities. The expression of genes for reactive oxygen-scavenging enzymes in maize rises with arsenic exposure [74].

Arsenic sensitivity varies greatly among different plant species. Arsenic accumulates to varied degrees in crops that receive irrigation water poisoned with arsenic, depending on the species and variety of the crop. Photosynthesis is among the most crucial physiological traits of plants, along with other metabolic processes. Research has shown that arsenic's impact on photosynthesis can reduce plants' capacity to produce food [75]. Arsenic's mobility and bioavailability may be influenced by the amount and kind of organic matter in the soil [60]. Plants exhibit morphological and physio-biochemical aberrations as a result of arsenic toxicity. It has been asserted that arsenic prevents root growth and expansion since the root is the organ that is most frequently exposed to the poison. Arsenic may also harm plants by resulting in stunted roots, withered leaves, a decline in the pigment required for photosynthetic activity, yellowing of the leaves, and a drop in chlorophyll, which can affect plant metabolism. However, because some of the defense mechanisms have not yet developed, seed germination and the early phases of seedling growth are more vulnerable to metal pollution and are therefore important considerations in the assessment of toxicity [76]. Arsenic binds to enzymes and proteins that harm cell biochemistry in addition to interfering with physiological activities such as photosynthesis, respiration, and transpiration in plants. It can lessen a plant's capacity for reproduction, obstructing the photosynthetic process, which lowers plant growth and output. While the entire plant in Arum is found to contain a significant concentration of arsenic [62]. Frequently, natural soil-grown plants contain little or no arsenic (3.6 mg/kg). *Pterisvittata* is a hyperaccumulator of arsenic, producing up to six times as much of the metal as is present in soil [70]. While arsenic in rice straw was shown to be favorably correlated with soil arsenic, arsenic in rice grain was found to be inversely correlated with soil arsenic, with arsenic in rice grain being 0.54 mg/kg in soil [72]. This suggests that rice can absorb a significant amount of As even at concentrations that are significantly lower than those that are dangerous. When irrigation water's concentration of arsenate was increased, plant height, grain yield, the number of filled grains, grain weight, and root biomass all drastically decreased. However, root, straw, and rice husk arsenic concentrations rose dramatically. The amount of arsenic in rice grain did not go beyond the acceptable level for food hygiene (1.0 mg of As kg^{-1} dry weight). As the levels of arsenic in rice straw (up to 91.8 mg/kg for the highest As treatment) were comparable to those in roots (up to 107.5 mg/kg), arsenic can travel quickly from the root to the shoot [77]. Even at the highest dose, the Boro variety V8 (Purbachi) showed excellent arsenate tolerance (8 mg As per l). Rice seeds typically germinate successfully up to 2 mg As per l, regardless of the type of arsenic. The root tolerance index (RTI) and relative shoot height (RSH) of rice seedlings decreased with increasing concentrations of arsenite and arsenate, according to a study on the short-term toxicity of arsenic [78].

The root membranes of tomato plants (Lycopersicumesculentum) were shown to be destroyed at 10 mg/l of arsenic in an experiment conducted by Barrachina et al. [79]. Exposure to arsenic significantly decreased key plant development indicators, with the largest decreases occurring at 76.8% for leaf fresh weight and 79.6% for tomato fruit output, respectively. Smith et al. evaluated the body of research and found that whereas maize and radish have a value of 100 mg/kg, rice, beans, and oats only have a value of 20 mg/kg. Arsenic is translocated to the grain and shoots in lentils, where it can significantly impede physiological development by delaying or preventing the creation of biomass and hindering plant reproduction. Stress also affects lipid peroxidation, electrolyte leakage, H2O2 accumulation, root oxidizability, and antioxidant enzyme activity in a substantial way [80]. When soil arsenic concentrations increase, so does the amount of arsenic that plants deposit. Terrestrial plants, such as legume crops, however, have a higher concentration of arsenic in the shoot-to-root region as compared to emergent plants [81], leading to physiological disorders in plants.

9. Conclusion

The toxicity of arsenic in numerous crops is a severe problem for human health. As exposure often deregulates several cellular functions, including epigenetic control, DNA repair, apoptosis resistance, and regular gene expression. Targeting these signaling pathways may offer a therapeutic method or alternative for the treatment and prevention of malignancies linked to chronic arsenic exposure because of their importance in the progression of carcinogenesis. When plants are exposed to arsenic, it induces reactive oxygen species (ROS), which put the plants under oxidative stress, eventually causing damage not only at the cellular level but also at the genetic level. Thus, the need of the hour is to develop an effective and cost-effective method to combat the accumulation of arsenic in different parts of the plants.

Author details

Sweety Nath Barbhuiya[1,2]*, Ringthardi Warisa[3], Dharmeswar Barhoi[3], Jurimoni Das[3] and Sarbani Giri[1]

1 Laboratory of Cell and Molecular Biology, Department of Life Science and Bioinformatics, Assam University, Silchar, Assam, India

2 Department of Zoology, Patharkandi College, Karimganj, Assam, India

3 Department of Zoology, The Assam Royal Global University, Guwahati, Assam, India

*Address all correspondence to: sweetynath91@gmail.com

References

[1] Singh R, Singh S, Parihar P, Singh VP, Prasad SM. Arsenic contamination, consequences, and remediation techniques: a review. Ecotoxicology and Environmental Safety. 2015;**112**:247-270

[2] Barbhuiya SN, Barhoi D, Giri S. Application of nanotechnology for heavy metals remediation from contaminated water. In: Shukla SK, Kumar S, Madhav S, Mishra SK, editors. Metals in Water. Massachusetts, USA: Elsevier; 2023. pp. 369-386

[3] Hettick BE, Canas-Carrell JE, French AD, Klein DM. Arsenic: a review of the element's toxicity, plant interactions, and potential methods of remediation. Journal of Agricultural and Food Chemistry. 2015;**63**(32):7097-7107

[4] Argos M, Kalra T, Rathouz PJ, Chen Y, Pierce B, Parvez F, et al. Arsenic exposure from drinking water, and all-cause and chronic-disease mortalities in Bangladesh (HEALS): a prospective cohort study. The Lancet. 2010;**376**(9737):252-258

[5] Basu A, Saha D, Saha R, Ghosh T, Saha B. A review on sources, toxicity and remediation technologies for removing arsenic from drinking water. Research on Chemical Intermediates. 2014;**40**:447-485

[6] Nath Barbhuiya S, Barhoi D, Giri A, Giri S. Arsenic and smokeless tobacco exposure induces DNA damage and oxidative stress in reproductive organs of female Swiss albino mice. Journal of Environmental Science and Health, Part C. 2020;**38**(4):384-408

[7] Barbhuiya SN, Barhoi D, Giri S. Impact of Arsenic on Reproductive Health. In: Otsuki T, editor. Environmental Health. London, UK: IntechOpen; 2021. p. 89-102

[8] Mondal S, Mukherjee S, Chaudhuri K, Kabir SN, Kumar Mukhopadhyay P. Prevention of arsenic-mediated reproductive toxicity in adult female rats by high protein diet. Pharmaceutical Biology. 2013;**51**(11):1363-1371

[9] Das S, Upadhaya P, Barhoi D, Nath Barbhuiya S, Langthasa P, Giri S. GCMS analysis of sadagura (smokeless tobacco), its enhanced genomic instability causing potential due to arsenic co-exposure, and vitamin-C supplementation as a possible remedial measure: a study involving multiple model test systems. Drug and Chemical Toxicology. 2022;**45**(1):185-196

[10] Ratnaike RN. Acute and chronic arsenic toxicity. Postgraduate Medical Journal. 2003;**79**(933):391-396

[11] Martinez VD, Becker-Santos DD, Vucic EA, Lam S, Lam WL. Induction of human squamous cell-type carcinomas by arsenic. Journal of Skin Cancer. 2011. pp. 1-9

[12] Meharg AA, Williams PN, Adomako E, Lawgali YY, Deacon C, Villada A, et al. Geographical variation in total and inorganic arsenic content of polished (white) rice. Environmental Science & Technology. 2009;**43**(5):1612-1617

[13] Smedley PL, Kinniburgh DG. A review of the source, behaviour and distribution of arsenic in natural waters. Applied Geochemistry. 2002;**17**(5):517-568

[14] Beniwal R, Yadav R, Ramakrishna W. Multifarious effects of arsenic on plants and strategies for mitigation. Agriculture. 2023;**13**(2):401

[15] Quaghebeur M, Rengel Z. Arsenic speciation governs arsenic uptake

and transport in terrestrial plants. Microchimica Acta. 2005;**151**:141-152

[16] Mohan D, Pittman CU Jr. Arsenic removal from water/wastewater using adsorbents—a critical review. Journal of Hazardous Materials. 2007;**142**(1-2):1-53

[17] Geng CN, Zhu YG, Hu Y, Williams P, Meharg AA. Arsenate causes differential acute toxicity to two P-deprived genotypes of rice seedlings (Oryzasativa L.). Plant and Soil. 2006;**279**:297-306

[18] Zhao FJ, Ma JF, Meharg AA, McGrath SP. Arsenic uptake and metabolism in plants. New Phytologist. 2009;**181**(4):777-794

[19] Nearing MM, Koch I, Reimer KJ. Complementary arsenic speciation methods: A review. Spetrochimica Acta Part B: Atomic Spectroscopy. 2014;**99**:150-162

[20] Gaus I, Kinniburgh DG, Talbot JC, Webster R. Geostatistical analysis of arsenic concentration in groundwater in Bangladesh using disjunctive kriging. Environmental Geology. 2003;**44**:939-948

[21] Smith AH, Lingas EO, Rahman M. Contamination of drinking-water by arsenic in Bangladesh: a public health emergency. Bulletin of the World Health Osrganization. 2000;**78**(9):1093-1103

[22] Ravenscroft P, Brammer H, Richards K. Arsenic in North America and Europe. In: Ravenscroft P, Brammer H, Richards K, editors. Arsenic Pollution: a Global Synthesis. New Jersey, USA: Wiley; 2009. pp. 387-454

[23] Das S, Langthasa P, Barhoi D, Upadhaya P, Giri S. Effect of nutritional status on arsenic and smokeless tobacco induced genotoxicity, sperm abnormality and oxidative stress in mice in vivo. Environmental and Molecular Mutagenesis. 2018;**59**(5):386-400

[24] Williams PN, Islam MR, Adomako EE, Raab A, Hossain SA, Zhu YG, et al. Increase in rice grain arsenic for regions of Bangladesh irrigating paddies with elevated arsenic in groundwaters. Environmental Science & Technology. 2006;**40**(16):4903-4908

[25] Meharg AA, Zhao FJ, Meharg AA, Zhao FJ. Arsenic in Rice Grain. Netherlands: Springer; 2012

[26] Agency for Toxic Substances and Disease Registry (ATSDR). (2007). Toxicological profile for wood creosote, coal tar creosote, coal tar, coal tar pitch, and coal tar pitch volatiles. U.S. Department of Health and Human Services, Public Health Service

[27] Abdul KS, Jayasinghe SS, Chandana EP, Jayasumana C, De Silva PM. Arsenic and human health effects: a review. Environmental Toxicology and Pharmacology. 2015;**40**(3):828-846

[28] Kumar A, Ali M, Kumar R, Kumar M, Sagar P, Pandey RK, et al. Arsenic exposure in Indo Gangetic plains of Bihar causing increased cancer risk. Scientific Reports. 2021;**11**:2376

[29] Moon KA, Guallar E, Umans JG, Devereux RB, Best LG, Francesconi KA, et al. Association between exposure to low to moderate arsenic levels and incident cardiovascular disease: a prospective cohort study. Annals of Internal Medicine. 2013;**159**(10):649-659

[30] Engel RR, Smith AH. Arsenic in drinking water and mortality from vascular disease: an ecologic analysis in 30 counties in the United States. Archives of Environmental

Health: An International Journal. 1994;**49**(5):418-427

[31] Wang SL, Chiou JM, Chen CJ, Tseng CH, Chou WL, Wang CC, et al. Prevalence of non-insulin-dependent diabetes mellitus and related vascular diseases in southwestern arseniasis-endemic and nonendemic areas in Taiwan. Environmental Health Perspectives. 2003;**111**(2):155-159

[32] Hassan FI, Niaz K, Khan F, Maqbool F, Abdollahi M. The relation between rice consumption, arsenic contamination, and prevalence of diabetes in South Asia. EXCLI Journal. 2017;**16**:1132

[33] Singh N, Kumar D, Sahu AP. Arsenic in the environment: effects on human health and possible prevention. Journal of Environmental Biology. 2007;**28**(2):359

[34] Mukherjee SC, Rahman MM, Chowdhury UK, Sengupta MK, Lodh D, Chanda CR, et al. Neuropathy in arsenic toxicity from groundwater arsenic contamination in West Bengal, India. Journal of Environmental Science and Health, Part A. 2003;**38**(1):165-183

[35] Dauphiné DC, Ferreccio C, Guntur S, Yuan Y, Hammond SK, Balmes J, et al. Lung function in adults following in utero and childhood exposure to arsenic in drinking water: preliminary findings. International Archives of Occupational and Environmental Health. 2011;**84**:591-600

[36] Saha JC, Dikshit AK, Bandyopadhyay M, Saha KC. A review of arsenic poisoning and its effects on human health. Critical Reviews in Environmental Science and Technology. 1999;**29**(3):281-313

[37] Järup L, Pershagen G, Wall S. Cumulative arsenic exposure and lung cancer in smelter workers: a dose-response study. American Journal of Industrial Medicine. 1989;**15**(1):31-41

[38] Welch K, Higgins I, Oh M, Burchfiel C. Arsenic exposure, smoking, and respiratory cancer in copper smelter workers. Archives of Environmental Health: An International Journal. 1982;**37**(6):325-335

[39] Kapaj S, Peterson H, Liber K, Bhattacharya P. Human health effects from chronic arsenic poisoning–a review. Journal of Environmental Science and Health, Part A. 2006;**41**(10):2399-2428

[40] Sasaki A, Oshima Y, Fujimura A. An approach to elucidate potential mechanism of renal toxicity of arsenic trioxide. Experimental Hematology. 2007;**35**(2):252-262

[41] Singh S, Yadav R, Sharma S. Singh AN. Arsenic contamination in the food chain: a threat to food security and human health. Journal of Applied Biology & Biotechnology. 2023;**10**(20):1-10

[42] Sarkar SD, Swain PR, Manna SK, Samanta S, Majhi P, Bera AK, et al. Arsenic contamination in food chain-a menace to food safety, human nutrition and health. Journal of Environmental Biology. 2022;**43**(3):339-349

[43] Peralta-Videa JR, Lopez ML, Narayan M, Saupe G, Gardea-Torresdey J. The biochemistry of environmental heavy metal uptake by plants: implications for the food chain. The International Journal of Biochemistry & Cell Biology. 2009;**41**(8-9):1665-1677

[44] Haque MM, Niloy NM, Khirul MA, Alam MF, Tareq SM. Appraisal of probabilistic human health risks of heavy metals in vegetables from industrial, non-industrial and arsenic contaminated

areas of Bangladesh. Heliyon. 2021;7(2):e06309

[45] Kazi TG, Brahman KD, Afridi HI, Arain MB, Talpur FN, Akhtar A. The effects of arsenic contaminated drinking water of livestock on its total levels in milk samples of different cattle: risk assessment in children. Chemosphere. 2016;**165**:427-433

[46] Mottalib MA, Zilani G, Suman TI, Ahmed T, Islam S. Assessment of trace metals in consumer chickens in Bangladesh. Journal of Health and Pollution. 2018;**8**(20):181208

[47] Le XC, Lu X, Li XF. Peer reviewed: arsenic speciation. Analytical Chemistry. 2004;**1**:26-33

[48] Bentley K, Soebandrio A. Arsenic and mercury concentrations in marine fish sourced from local fishermen and fish markets in mine-impacted communities in Ratatotok Sub-district, North Sulawesi, Indonesia. Marine Pollution Bulletin. 2017;**120**(1-2):75-81

[49] Tripathi RD, Tripathi P, Dwivedi S, Dubey S, Chatterjee S, Chakrabarty D, et al. Arsenomics: omics of arsenic metabolism in plants. Frontiers in Physiology. 2012;**3**:275

[50] Finnegan PM, Chen W. Arsenic toxicity: the effects on plant metabolism. Frontiers in Physiology. 2012;**3**:182

[51] Hughes MF. Arsenic toxicity and potential mechanisms of action. Toxicology Letters. 2002;**133**(1):1-6

[52] Martínez-Castillo JI, Saldaña-Robles A, Ozuna C. Arsenic stress in plants: A metabolomic perspective. Plant. Stress. 2022;**3**:100055

[53] Pickering IJ, Prince RC, George MJ, Smith RD, George GN, Salt DE. Reduction and coordination of arsenic in Indian mustard. Plant Physiology. 2000;**122**(4):1171-1178

[54] Abbas G, Murtaza B, Bibi I, Shahid M, Niazi NK, Khan MI, et al. Arsenic uptake, toxicity, detoxification, and speciation in plants: physiological, biochemical, and molecular aspects. International Journal of Environmental Research and Public Health. 2018;**15**(1):59

[55] Kumar K, Gupta D, Mosa KA, Ramamoorthy K, Sharma P. Arsenic transport, metabolism, and possible mitigation strategies in plants. In: Srivastava S, Srivastava AK, Suprasanna P, editors. Plant- Metal Interactions. Swizerland AG: Springer Nature; 2019:141-168

[56] Zhang J, Hamza A, Xie Z, Hussain S, Brestic M, Tahir MA, et al. Arsenic transport and interaction with plant metabolism: Clues for improving agricultural productivity and food safety. Environmental Pollution. 2021;**290**:117987

[57] Garg N, Singla P. Arsenic toxicity in crop plants: physiological effects and tolerance mechanisms. Environmental Chemistry Letters. 2011;**9**:303-321

[58] Meharg AA, Hartley-Whitaker J. Arsenic uptake and metabolism in arsenic resistant and nonresistant plant species. New Phytologist. 2002;**154**(1):29-43

[59] Farooq MA, Islam F, Ali B, Najeeb U, Mao B, Gill RA, et al. Arsenic toxicity in plants: cellular and molecular mechanisms of its transport and metabolism. Environmental and Experimental Botany. 2016;**132**:42-52

[60] Anawar HM, García-Sánchez A, Zabed HM. Biogeochemical cycling

of arsenic in soil–plant continuum: perspectives for phytoremediation. In: Gupta DK, Corpas FJ, Palma JM, editors. Heavy Metal Stress in Plants. Swizerland AG: Springer Nature; 2013:203-224

[61] Gusman GS, Oliveira JA, Farnese FS, Cambraia J. Arsenate and arsenite: the toxic effects on photosynthesis and growth of lettuce plants. Acta Physiologiae Plantarum. 2013;**35**:1201-1209

[62] Huq SI, Joardar JC, Parvin S, Correll R, Naidu R. Arsenic contamination in food-chain: transfer of arsenic into food materials through groundwater irrigation. Journal of Health, Population, and Nutrition. 2006;**24**(3):305

[63] Fitz WJ, Wenzel WW. Arsenic transformations in the soil–rhizosphere–plant system: fundamentals and potential application to phytoremediation. Journal of Biotechnology. 2002;**99**(3):259-278

[64] Mishra BK, Dubey CS, Shukla DP, Bhattacharya P, Usham AL. Concentration of arsenic by selected vegetables cultivated in the Yamuna flood plains (YFP) of Delhi, India. Environmental Earth Sciences. 2014;**72**:3281-3291

[65] Wu X, Hu J, Wu F, Zhang X, Wang B, Yang Y, et al. Application of TiO_2 nanoparticles to reduce bioaccumulation of arsenic in rice seedlings (Oryzasativa L.): a mechanistic study. Journal of Hazardous Materials. 2021;**405**:124047

[66] Mawia AM, Hui S, Zhou L, Li H, Tabassum J, Lai C, et al. Inorganic arsenic toxicity and alleviation strategies in rice. Journal of Hazardous Materials. 2021;**408**:124751

[67] Marin AR, Masscheleyn PH, Patrick WH. Soil redox-pH stability of arsenic species and its influence on arsenic uptake by rice. Plant and Soil. 1993;**152**:245-253

[68] Xie ZM, Huang CY. Control of arsenic toxicity in rice plants grown on an arsenic-polluted paddy soil. Communications in Soil Science and Plant Analysis. 1998;**29**(15-16):2471-2477

[69] Marin AR, Masscheleyn PH, Patrick WH. The influence of chemical form and concentration of arsenic on rice growth and tissue arsenic concentration. Plant and Soil. 1992;**139**:175-183

[70] Ma LQ, Komar KM, Tu C, Zhang W, Cai Y, Kennelley ED. A fern that hyperaccumulates arsenic. Nature. 2001;**409**(6820):579

[71] Ding D, Li W, Song G, Qi H, Liu J, Tang J. Identification of QTLs for arsenic accumulation in maize (Zea mays L.) using a RIL population. PLoS One. 2011;**6**(10):e25646

[72] Gupta A, Dubey P, Kumar M, Roy A, Sharma D, Khan MM, et al. Consequences of arsenic contamination on plants and mycoremediation-mediated arsenic stress tolerance for sustainable agriculture. Plants. 2022;**11**(23):3220

[73] Roy P, Saha A. Metabolism and toxicity of arsenic: a human carcinogen. Current Science. 2002;**82**(1):38-45

[74] Mylona PV, Polidoros AN, Scandalios JG. Modulation of antioxidant responses by arsenic in maize. Free Radical Biology and Medicine. 1998;**25**(4-5):576-585

[75] Zemanová V, Pavlíková D, Hnilička F, Pavlík M. Arsenic toxicity-induced physiological and metabolic changes in the shoots of Pteriscretica and Spinaciaoleracea. Plants. 2021;**10**(10):2009

[76] Liu X, Zhang S, Shan X, Zhu YG. Toxicity of arsenate and arsenite on germination, seedling growth and amyl olytic activity of wheat. Chemosphere. 2005;**61**(2):293-301

[77] Abedin MJ, Cresser MS, Meharg AA, Feldmann J, Cotter-Howells J. Arsenic accumulation and metabolism in rice (Oryzasativa L.). Environmental Science & Technology. 2002;**36**(5):962-968

[78] Abedin MJ, Meharg AA. Relative toxicity of arsenite and arsenate on germination and early seedling growth of rice (Oryzasativa L.). Plant and Soil. 2002;**243**:57-66

[79] Barrachina AC, Carbonell FB, Beneyto JM. Arsenic uptake, distribution, and accumulation in tomato plants: effect of arsenite on plant growth and yield. Journal of Plant Nutrition. 1995;**18**(6):1237-1250

[80] Smith ER, Naidu R, Alston AM. Arsenic in the soil environment. In: Sparks DL, editor. Advances in Agronomy. Massachusetts, USA: Academic Press; 1998;**64**:149-189

[81] Alam MZ, Hoque MA, Ahammed GJ, McGee R, Carpenter-Boggs L. Arsenic accumulation in lentil (Lens culinaris) genotypes and risk associated with the consumption of grains. Scientific Reports. 2019;**9**(1):9431

Chapter 4

Arsenic Exposure and Health Effects: Differences by Chemical Structure, Chemical Form and Arsenic Methylation Capacity

Hiroshi Yamauchi, Takahiko Yoshida and Ayako Takata

Abstract

Large-scale chronic arsenic poisoning in several areas worldwide has not been eradicated. Adverse health effects of inorganic arsenic (iAs) exposure are associated with decreased arsenic methylation capacity (AMC). The AMC results from 21 children and 39 adults with acute arsenic poisoning support and effectively guide the analysis and evaluation of AMC in chronic arsenic poisoning. Results of the 65th-year follow-up for 6223 cases of infants, who survived subacute arsenic poisoning, contribute to future projections against health effects (cancer and cognitive dysfunction). Currently, arsenolipids from fish and shellfish are attracting attention because of their direct association with brain dysfunction. Accordingly, the European Food Safety Authority is considering legal restrictions on their dietary intake. Modern society overlooks the environmental contamination and health hazards caused by arsenic chemical weapons mainly produced during World War II. This review aims to evaluate the various health effects of AMC reduction in acute, subacute, and chronic arsenic poisoning, in addition to the methylarsenic compounds such as arsenolipids, arsenosugars, and arsenobetaine, as well as arsenic chemical weapons such as lewisite, diphenylchloroarsin, and diphenylcyanoarsine.

Keywords: arsenic, arsenic poisoning, arsenic methylation capacity, arsenolipids, arsenic chemical weapons, infants, cognitive abilities, cognitive dysfunction

1. Introduction

As a result of naturally occurring inorganic arsenic (iAs) leached into aquifers and polluted drinking water, large-scale chronic arsenic poisoning in various areas of the worldwide occurred and has not ended even today. As the source of iAs contamination cannot be fundamentally removed, chronic arsenic poisoning is difficult to eradicate. The adverse health effects of arsenic exposures observed on the general population from the fetus through maternal placenta to the elderly comprise lifestyle disorders and cognitive dysfunctions in addition to the conventional poisoning symptoms such as skin lesions and malignancy. The health effects of iAs exposure are expected to be strongly associated with a reduction in arsenic methylation capacity (AMC) [1, 2]. In general, various confounding factors can affect the AMC results obtained from urinary samples. Valid supportive information on this issue is based on the occurrence of acute arsenic poisoning in 61 patients [3]; however, it virtually included no confounding effect and was able to serve as a reference value in the AMC evaluation in chronic arsenic poisoning. The results of a survey of 6223 survivors at the time of 65 years of follow-up among the 12,131 subjects exposed to arsenic sub-acutely in infancy [4] provide important data in predicting effects on the future health outcomes of chronic arsenic exposure, especially on carcinogenesis and the cognitive impairment resulting from the damage to the brain and central nervous system.

Conventionally, the complications of arsenic exposure are attributed only to iAs. The seafood for daily consumption contains arsenolipids (AsLipids) with an estimated stronger toxicity than iAs [5, 6]. Accordingly, the European Food Safety Agency (EFSA) is preparing a legal regulation focusing on AsLipids ingested via seafood [7]. On the other hand, arsenic chemical weapons, largely produced during World War II, have been causing environmental contamination due to abandoned phenylarsenic chemical warfare agents in the Baltic Sea [8–10], Germany [11], China [12–14], and Japan [15]. The degradation product of phenylarsenic chemical warfare agents is diphenylarsinic acid (DPAA). In Japan, poisoning by DPAA has been confirmed in 157 patients [15]. The target organ of this DPAA is the brain, where it can infiltrate the blood-brain barrier (BBB) and cause cerebral central nervous system damage.

This review aims to outline the differences in health effects of acute, subacute, and chronic arsenic poisoning based on the results of the AMC. In addition, it presents the adverse effects of AsLipids and phenylarsenic chemical warfare agents, which are likely to be more toxic than iAs.

2. Factors affecting the occurrence and the symptoms of acute arsenic poisoning

2.1 Crystallized and dissolved arsenic trioxide (ATO)

Arsenic compounds have long been used in many countries for intentional suicide and homicide. The background of their utilization is related to their physical properties of being tasteless, odorless, and non-irritating. Powdered arsenic trioxide (ATO) has been used in many cases of suicide (**Table 1**) [16–20] and by some medical personnel with access to ATO [17]. Factors such as the arsenic dose ingested, the form of ATO (crystals or solution), and time to life-saving treatment play a key role in the development of acute arsenic poisoning. It should be noted that in case reports of acute arsenic poisoning, most references to ATO intake are in terms of the swallowed weight (g) rather than the actual amount absorbed into the body. ATO crystals do not rapidly dissolve in water, predicting their gradual dissolution in the gastrointestinal tract; this slow dissolution rate provides an opportunity for emergency treatment, such as washing the stomach and the intestines, and adsorption and removing ATO with activated carbon. As ATO is a metalloid with metallic properties, an abdominal X-ray of the patient should show a white shadow on the wall of the gastrointestinal tract [16, 19, 20], indicating the ingestion status. A critically ill patient who ingested a large amount of ATO (54 g) was successfully saved by surgically removing the ATO through an incision in the stomach [19]. In other words, in acute arsenic poisoning

Country	As intake (g)[a]	As level in urine and blood	Time (hr)[b]	Chelating agent	Days to discharge	Ref.
United Kingdom[a]	75	Blood 2300 μg/l	4	DMSA	Death, 0.7	[16]
Japan[a]	1.04	Urine, 51,400 μg/l[c] Blood, 415 μg/l[c] Day 14 Urine, 900 Blood, 70	20	BAL	26	[17]
Germany[a]	0.6	Urine, μg/l Day 1, 215,500 Day 2, 6049 Day 8, 169	7	DMPA	12	[18]
Spain[a]	54	Urine, μg/l Day 1, 67,500 Day 10, 3000 Blood, μg/l Day 1, 132 Day 10, 15	5	BAL D-Peni- cillamine	55	[19]
Taiwan[a]	37.5	Urine, μg/g.cr. 284,576[c] Blood, μg/l 2160[c] Day 1 Urine, 19,583 Blood, 417.5 Day 5 Urine, 1475 Blood, 38.9	1.5	DMPA	22	[20]
Japan[d]	Children n = 22 64.5 (48.3–80.5) mg[e] Adults n = 39 76.0 (56.0–80.5) mg[e]	Urine, μg/g.cr.[e] Day 1 Children, n = 4 14,273 (9138 – 20,346) Adults, n = 6 5251 (3602 – 7418)	< 1	non	4 deaths 0.5 Children, 7 Adults, 30–60	[3]

[a]*Acute arsenic poisoning caused by ingestion of crystalline (powdered) ATO.*
[b]*Time from ATO ingestion to initiation of treatment.*
[c]*Value at hospitalization (before chelating agent administered).*
[d]*Acute arsenic poisoning caused by ingestion of dissolved ATO.*
[e]*Values are median (IQR).*
DMSA, 2, 3-dimercaptosuccinic acid; DMPA, 2,3-dimercapto-1-propanesulfonic acid;
BAL, 2,3-dimercaptopropanol.

Table 1.
Information on the incidence requirements and chelating agents for acute arsenic poisoning due to arsenic trioxide (ATO).

caused by crystalline ATO, the removal of ATO residues from the gastrointestinal tract has been shown to be extremely important for the patient's survival.

An acute arsenic poisoning of 61 people, including children (n = 22) and adults (n = 39), in Wakayama, Japan, was caused by the intentional addition of industrial

ATO to hot curry soup. Later investigations revealed that most of the ATO was dissolved (with small amounts of crystals) in the residual heat of the curry soup. The estimated arsenic intake amount (body absorption) was calculated using the arsenic in the patient's urine. The arsenic doses in children and adults were 67.4 ± 29.5 mg and 78.3 ± 38.7 mg, respectively. Information on the lethal dose in humans is scarce; however, it is nearly 300 mg [21]. Four people (aged 10, 16, 53, and 64 years) died by approximately 12 h after ingesting ATO. Dissolved ATO is presumed to be absorbed faster and at a higher rate in the gastrointestinal tract than crystallized ATO. Adults with severe symptoms have taken 100–200 mg of ATO. In contrast, even when the ingested arsenic dose exceeded 100 mg, the children had mild symptoms and were discharged from the hospital after about 1 week. The differences in symptoms between children and adults are discussed in later chapters (see 2.3).

In all patients, the ingestion of dissolved ATO caused severe nausea and vomiting within approximately 5 min [3]. Gastrointestinal symptoms were observed even at the minimum body arsenic dose of 22 mg (1-year-old infant). Afterward, watery diarrhea developed in approximately 30% of the patients. Laboratory values showed an initial increase in white blood cell count, followed by a rapid decrease to reach a minimum on days 3–6. The liver dysfunction (abnormal increase of GOT and GPT) peaked at about 1 week and then slowly recovered. The prolongation of the Q–T interval was characteristic of the electrocardiogram and was also observed in children. In adults with severe disease, erythematous macule, without pain or itching, was found as skin lesions on the abdomen, armpits, and neck [22]. In addition, peripheral neuropathy (paresthesia and motor paralysis) was observed [23]. Patients showed white shadows on the wall of their gastrointestinal tract on the abdominal X-ray, consistent with the observations on the reported victims who ingested crystalline ATO [16, 19, 20]. Therefore, the symptoms of acute arsenic poisoning by dissolved ATO tend to develop quickly and severely ill, especially in adults.

2.2 Effects of chelating agent treatment

British Anti-Lewisite (BAL), initially developed against the arsenic chemical weapon-lewisite, was subsequently used to treat acute iAs poisoning. However, BAL has disadvantages, such as severe pain with muscle injection. The use of chelating agents in acute arsenic poisoning is to promote urinary excretion by binding iAs^{III} to SH groups. To achieve a high rate of this pharmacological reaction, iAs^{III} should be not oxidized or methylated to monomethylarsonic acid (MMA) or dimethylarsinic acid (DMA), and the chelating agent must be used as soon as possible after ATO ingestion. This mechanism of action is common to BAL, 2,3-dimercapto-1-propanesulfonic acid (DMPA) and 2, 3-dimercaptosuccinic acid (DMSA) [24, 25]. DMPA tends to be used orally or intramuscularly, depending on the patient's symptoms [20].

On the other hand, DMPA or DMSA as a treatment for chronic arsenic poisoning due to iAs exposure has been tested in India and Bangladesh without efficacy [24, 25]. However, studies are still ongoing for the development of an effective therapeutic drug for chronic arsenic poisoning.

2.3 Differences in symptoms between children and adults: evaluation by AMC

Epidemiological studies of chronic arsenic poisoning have been attempted to elucidate the relationship between health effects and AMC. In summary, the studies suggested that any decrease in the metabolic efficiency of methylation to iAs can lead to health deterioration. The indicators of an AMC decrease are an increase in %MMA and a decrease in the second methylation ratio (DMA/MMA) or the secondary methylation index (SMI; DMA/MMA + DMA). Recent reports indicate a trend toward higher rates of AMC in children compared to adults [26–28]. Chronic arsenic poisoning occurs more often in economically developing countries, and the situation of poisoning has been related to the environment, diet, and hygienic conditions in which the subjects live. There are various confounding factors in the urinary arsenic results obtained from subjects, especially the influence of nutrients [29, 30].

To our knowledge, there are no studies evaluating AMC in patients with acute arsenic poisoning, including ATO. The 61 patients with acute arsenic poisoning cases presented in this review did not have the opportunity for emergency treatment with chelating agents [3]. Therefore, their urinary arsenic results reflected only the observed results without confounding factors such as nutrients and arsenic species in the diet, which are usually noted in epidemiological studies of chronic arsenic poisoning. Although the estimated arsenic intake amount of ATO was similar among the patients (see section 2.1), the symptoms were milder in children and, conversely, more significant in adults [3]. The characteristics of AMC in children and adults are shown in **Figure 1**. Children had lower %MMA and higher second methylation ratio or SMI. In adults with severe symptoms, we observed a significant trend toward increased %MMA and decreased second methylation ratio or SMI. In acute and chronic arsenic poisoning [27], the data available to elucidate the cause of high AMC

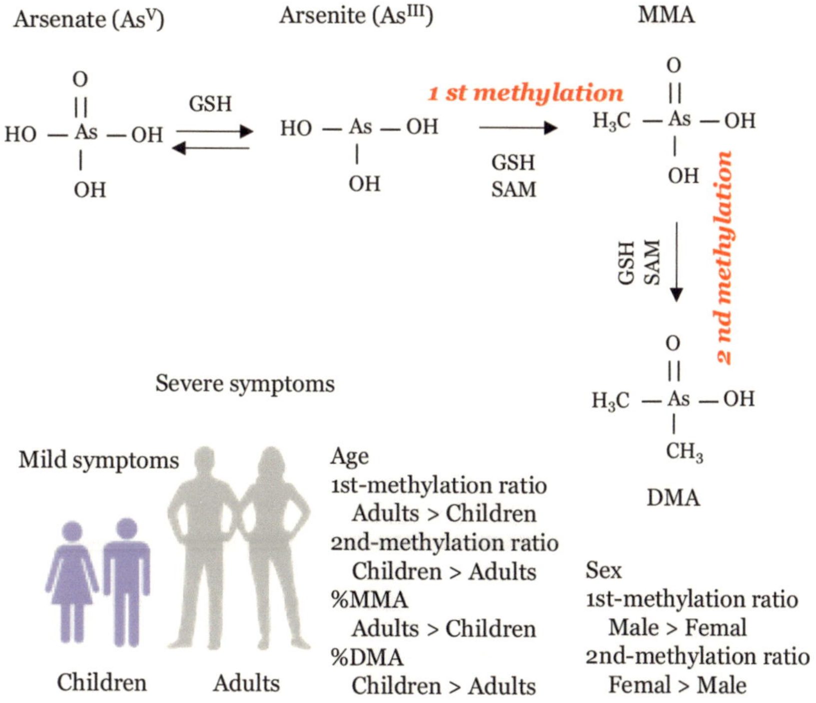

Figure 1.
Characteristics of arsenic methylation capacity (AMC) in children and adults with acute arsenic poisoning. Children are under 16 years old and adults are over 16 years old. Monomethylarsonic acid (MMA), dimethylarsinic acid (DMA), S-adenosylmethionine (SAM), reduced glutathione (GSH). First methylation ratio (MMA/inorganic arsenic (iAs)), second methylation ratio (DMA/MMA). Methylation efficiency was assessed by determining the relative amounts of total arsenic in urinary (%iAs, %MMA, and %DMA).

in younger children are limited. The mechanism that is currently receiving the most attention is its association with one-carbon metabolism (OCM). OCM is a biochemical pathway that provides methyl donor group S-adenosylmethionine (SAM) to methylate arsenic. It has been estimated that OCM is activated in younger children and degrades with age; there are several interesting reviews that have evaluated the link between OCM and arsenic methylation [1, 29, 31].

AMC findings from children and adults with acute arsenic poisoning may serve as a valid guideline for evaluating AMC in patients with chronic arsenic poisoning and those exposed to high concentrations of iAs.

3. After-effects of subacute arsenic poisoning in infants after 55–65 years

Subacute arsenic poisoning occurs under conditions of continuous intake of approximately 1/100th of the lethal dose of acute arsenic poisoning over several months. In 1955, out of 12,131 infants who developed subacute arsenic poisoning after ingesting the powdered Morinaga Milk contaminated with iAs (arsenate, iAs^{V}), 130 died. The contaminated milk was consumed on an average of 3 months with an estimated daily arsenic intake of 1.3–3.6 mg, leading to a total accumulation of 90–140 mg. The major initial symptoms were fever, vomiting, diarrhea, abdominal distension, hepatomegaly, cough, nasal discharge, conjunctivitis, and melanoderma. Nevertheless, the neurological indices might not have been properly identified due to the age of the patients who were infants. The confirmed laboratory results comprised anemia, electrocardiographic abnormalities, and a radiographic band-like shadow of the epiphysis of the long bone [4, 32].

The acute arsenic poisoning patients in Wakayama included four 1-year-old infants (mean ATO intake; 44 (22–65) mg) with mild symptoms [3]. In contrast, 130 infants of the same age died from subacute arsenic poisoning [4]. To date, this question has not been resolved, but we can speculate on possible interpretations on the basis of differences in AMC affected by exposure situation of iAs. The continual ingestion of iAs is assumed to accelerate the deterioration of physical condition. The iAs methylation requires the supply of the methyl group donor SAM and reduced glutathione (GSH) by the OCM pathway. If deteriorating physical conditions persist, the supply of SAM decreases due to the reduced function of the OCM [1, 29, 31]. It is assumed that the deterioration of physical conditions led to the depletion of SAM and GSH due to the inability of the nutritional supply to function, and as a result, the detoxification mechanism of iAs was blocked, leading to severe illness and even death.

Results of a subsequent follow-up survey in the 15th year showed a delayed growth, intellectual disability, central nervous system disorders such as epilepsy and hearing loss, and skin conditions such as melanoderma and keratosis [32]. Another survey in the 55th year demonstrated carcinogenesis and cognitive dysfunctions, with a significantly higher risk of mortality from neurological diseases compared to the general population [33]. Recently, the results of the 65th-year follow-up study were reported [4]. In total, 6223 survivors of the iAs poisoning were enrolled in a prospective cohort study from 1982 to 2018. Using standardized mortality rates (SMRs), the subjects' mortality and cancer incidence rates were compared to their respective Japanese population rates and 95% confidence intervals (95% CI) were calculated. A statistically significant increase in mortality ratio was observed for all causes (SMR, 1.15; 1.01–1.19), nervous system disease (2.83, 1.62–4.19), respiratory disease (2.02, 1.37–2.62), genitourinary systems disease (2.25, 1.10–3.73). It should be noted that epilepsy and cerebral palsy mainly contributed to the high mortality rate caused by the central nervous system disease. Therefore, it is suggested that subacute arsenic poisoning in infancy may cause permanent damage to the central nervous system. Furthermore, liver cancer showed a high mortality rate (SMR, 1.68, 1.06–2.31), whereas that of bladder cancer increased; however, no significant difference was detected. These findings are consistent with the epidemiological studies in Chile [34, 35] that suggested a link between exposure to high levels of iAs during infancy and the risk of developing bladder and lung cancers at an older age. Consequently, we consider that the results of the subsequent follow-up surveys applied to the subacute arsenic poisoning survivors are a warning sign against the risk of human exposure to iAs when the BBB is still not functional, such as in infants and fetuses.

4. Past, present, and expected future health effects from exposure to environmental iAs

In the early twentieth century, chronic arsenic poisoning was first discovered in Argentina, followed by large epidemics in Chile and Taiwan. Then, starting in the late 1970s, it was detected in India, Bangladesh, and China among Asian countries. Chronic arsenic poisoning occurred by consuming the provided drinking water without previous testing for the presence of toxic substances. Particularly well water triggered the outbreaks of serious poisoning. Because chronic arsenic poisoning is caused by naturally existing iAs and the source cannot be eliminated, the potential occurrence of epidemic to be continue in the future [32].

4.1 iAs exposure and skin lesions

Chronic arsenic poisoning generally develops from drinking well water contaminated with iAs^{III} or iAs^{V} (approximately 50 μg/L) over a period of more than several years. However, even low concentrations (5–10 μg/L) have been associated with the development of skin lesions [36]. The most prominent symptom of chronic arsenic poisoning is skin lesions, characterized by hyperkeratosis of the palms and soles and hyperpigmentation and depigmentation (raindrops) particularly pronounced on the abdomen and trunk. Hyperkeratosis is painful and can affect daily routine activities and reduce the quality of life.

The iAs exposure dose is the primary factor in the development of skin lesions. Another possible factor suggested factor is the decrease in AMC, which raises %MMA and decreases SMI [2]. This factor is more significant in males compared with females. On the other hand, skin lesions in children are significantly milder than in adults, suggesting that AMC in children is higher than in adults [27]. This is consistent with the acute arsenic poisoning case, in which children had milder symptoms than adults [3]. The mechanism between the decrease in AMC and skin lesions has not been clearly elucidated, but the genetic polymorphism of methylation has been explored [37, 38].

4.1.1 AMC determined from blood arsenic and evaluation of skin lesions

For unknown reasons, epidemiological studies of chronic arsenic poisoning have not reported the profile of arsenic species in the blood. The difficulty in analyzing arsenic in blood by chemical form (its species) may be one of the reasons for the lack of reports. In 1999, we measured the urinary and blood arsenic in a small sample (n = 78) in Gangfangying village, Inner Mongolia, China, without publishing the results. In this review, we reanalyzed the relationship between the index of AMC determined from blood arsenic and skin lesions.

The determination of the chemical form of arsenic in blood is similar to the method of urine analysis [39].

Of the 78 subjects [14 children (7–17 years) and 64 adults (21–70 years)], 38 had skin lesions (dyspigmentation and hyperkeratosis). **Table 2** shows the comparison of compares the urinary and blood arsenic concentrations and the index values of AMC (iAs, MMA, and DMA) in the subjects grouped by the presence and absence of skin lesions. Arsenic levels in urine and blood were significantly higher in the group with skin lesions ($p < 0.001$). However, urinary arsenic profile did not show any statistically significant difference in the AMC index, unlike blood arsenic profile ($p < 0.001$) (**Table 2**). In general, urinary arsenic is known to vary due to confounding factors. The urinary arsenic levels of the 78 subjects showed four cases of outliers for iAs, three for MMA, and another three for DMA, but no outliers for blood arsenic (results not shown in table). Tamayo et al. [40] noted that arsenic in blood has fewer outliers than arsenic in urine. The fact that AMC calculated based on blood arsenic is superior to that based on urinary arsenic in evaluation of the AMC of individual subjects is due to the result of the relatively stable of blood arsenic concentration.

Table 3 shows the adjusted odds ratios (OR) for age and sex for each of the variables listed in **Table 2**. In the AMC determined from of blood, %MMA showed a significant association with the risk of developing skin lesions (OR = 1.12, 95%CI: 1.06–1.19, $p < 0.001$). AMC determined from blood arsenic profiles was validated in a small group of subjects; however, more studies reaching various research areas are needed.

4.1.2 Improvement of skin lesions by reducing iAs exposure

We investigated through a one-year observation study whether the confirmed skin lesions of the 38 subjects could be ameliorated by reducing iAs doses from well water by changing water source. The average arsenic concentration in well water during iAs exposure was 0.13 ± 0.20 mg/l, while the one used for exposure reduction was 0.037 mg/l. The urinary IMD (iAs + MMA + DMA) concentration during iAs exposure was median (IQR) 200 (53.5–398) μg/g cr., while the IMD level at 1-year was median (IQR) 50.4 (33.3–83.0) μg/g cr. Dyspigmentation and hyperkeratosis changes are shown in **Figure 2**, where relative but insufficient improvement was observed. The AMC index determined from urinary arsenic in all 38 subjects did not significantly change. These patients were followed up for an additional 5, 10, and 15 years. A fast improvement in hyperkeratosis was observed in the first year, but, unexpectedly, successive improvement was found to be slow [41].

A study in Bangladesh suggested that reducing iAs exposure from well water (from >50 μg/l to <50 μg/l) over a 2-year period not only increased the recovery rate of skin lesions, but also reduced the severity of skin lesions in patients over time [42]. A study in Inner Mongolia, China, tested whether skin lesions improved by reducing the arsenic concentrations in the well water from >200 μg/l (2004) to <5 μg/l (2017). The 119 subjects were compared for skin symptoms (hyperkeratosis), urinary arsenic levels, and AMC indices in 2004 and 2017. As a result, 25 subjects (21%) showed improvement of skin symptoms. However, recovery of skin symptoms showed an association between %iAs and %DMA, no association between %MMA due to its wide variable value [43]. Reduction of iAs exposure from drinking water may not be sufficient for improvement of hyperkeratosis.

	No skin lesions (n = 40)	Skin lesions (n = 38)	p value
Age	35.5 (22.0–44.0)	40.0 (31.0–46.3)	0.188
Adults: Children	32: 8	33: 5	0.547[b]
Male: Female	14: 26	25: 13	0.012[b]
Cumulative arsenic dose[a]	0.092 (0.064–0.26)	1.63 (0.604–3.68)	< 0.001
U-IMD μg/g.cr.	53.6 (28.9–110)	262 (86.7–580)	< 0.001
U-iAs μg/g.cr.	8.94 (4.67–16.8)	40.7 (15.8–107)	< 0.001
U-MMA μg/g.cr.	4.11 (2.83–12.2)	33.2 (9.25–83.7)	< 0.001
U-DMA μg/g.cr.	43.7 (20.6–84.2)	200 (68.3–389)	< 0.001
U-1st methylation ratio	0.655 (0.424–0.874)	0.768 (0.602–0.937)	0.211
U-2nd methylation ratio	7.08 (5.53–11.5)	6.67 (3.57–8.40)	0.135
U-PMI	0.832 (0.795–0.874)	0.834 (0.796–0.877)	1.000
U-SMI	0.876 (0.847–0.920)	0.870 (0.781–0.894)	0.135
U-%iAs	16.8 (12.6–20.5)	16.6 (12.3–20.4)	1.000
U-%MMA	10.2 (6.79–12.9)	11.1 (8.50–17.0)	0.134
U-%DMA	71.7 (67.7–78.0)	73.0 (62.0–78.7)	0.478
B-IMD μg/l	4.05 (3.24–4.82)	8.77 (4.55–14.9)	< 0.001
B-iAs μg/l	2.00 (1.60–2.40)	3.20 (1.60–4.10)	< 0.001
B-MMA μg/l	0.89 (0.44–1.33)	2.89 (0.89–6.22)	< 0.001
B-DMA μg/l	1.14 (1.14–1.71)	2.29 (1.71–4.57)	< 0.001
B-1st methylation ratio	0.444 (0.288–0.667)	1.05 (0.556–1.85)	< 0.001
B-2nd methylation ratio	1.29 (1.29–2.57)	0.935 (0.643–1.29)	< 0.001
B-PMI	0.548 (0.458–0.653)	0.660 (0.598–0.765)	< 0.001
B-SMI	0.563 (0.563–0.720)	0.483 (0.391–0.563)	< 0.001
B-%iAs	45.2 (34.7–54.2)	34.0 (23.5–40.2)	< 0.001
B-%MMA	20.1 (17.0–27.5)	35.7 (24.2–41.9)	< 0.001
B-%DMA	33.3 (25.0–39.0)	33.2 (25.0–39.0)	0.621

[a]*μg/l years. Urine, U; blood, B. Values are median (IQR). IMD, inorganic arsenic (iAs) + methylarsonic acid (MMA) + dimethylarsinic acid (DMA). First methylation ratio (MMA/iAs), second methylation ratio (DMA/MMA). Primary methylation index (PMI, MMA + DMA/IMD), secondary methylation index (SMI, DMA/MMA + DMA).*
[b]*Mann-Whitney's U test was used for continuous variables, and Chi-square test was used for categorical variables to compare two groups.*

Table 2.
Comparison of urinary and blood arsenic concentrations and AMC in 78 residents of Gangfangying village, China.

4.2 iAs exposure and carcinogenesis

IARC has identified iAs as a Group 1 carcinogen that causes skin, lung, bladder, and liver cancers [44]. The latent period of arsenic carcinogenesis is generally estimated at 30–50 years. The data on carcinogenesis in patients with chronic arsenic

Variable	Adjusted OR (95% CI)	p value
Cumulative arsenic dose[a]	1.89 (1.25–2.85)	0.002
U-IMD μg/g.cr.	1.01 (1.00–1.02)	< 0.001
U-iAs μg/g.cr.	1.06 (1.02–1.09)	0.001
U-MMA μg/g.cr.	1.06 (1.02–1.11)	0.002
U-DMA μg/g.cr.	1.01 (1.01–1.02)	< 0.001
U-1st methylation ratio	2.09 (0.57–7.71)	0.266
U-2nd methylation ratio	0.99 (0.92–1.08)	0.897
U-PMI	2.52 (0.005–1370)	0.774
U-SMI	0.001 (o – 4.37)	0.109
U-%iAs	0.99 (0.93–1.10)	0.774
U-%MMA	1.08 (0.98–1.20)	0.121
U-%DMA	0.98 (0.93–1.04)	0.564
B-IMD μg/l	1.42 (1.17–1.74)	< 0.001
B-iAs μg/l	2.03 (1.32–3.10)	0.001
B-MMA μg/l	2.52 (1.44–4.40)	0.001
B-DMA μg/l	3.45 (1.76–6.77)	< 0.001
B-1st methylation ratio	10.3 (2.60–40.6)	< 0.001
B-2nd methylation ratio	0.49 (0.26–0.94)	0.033
B-PMI	2440 (22.5–263,800)	0.001
B-SMI	0.004 (0–0.26)	0.010
B-%iAs	0.92 (0.88–0.97)	0.001
B-%MMA	1.12 (1.06–1.19)	< 0.001
B-%DMA	1.01 (0.95–1.06)	0.838

[a]*μg/l years.*
Adjusted for age and sex. OR, odds ratio; CI, confidence interval.

Table 3.
OR and 95% CI for skin lesions by variable factors using arsenic methylation capacity calculated based on arsenic species in urine or blood.

poisoning recognize a trend toward reporting from countries with a relatively long history of iAs exposure. Evidence between iAs exposure and carcinogenesis has been reported from Taiwan for skin [45–47] and bladder cancers [48], and from Chile [49] and Argentina [50] for lung cancers. Previous reports have suggested a relationship between decreased AMC and carcinogenesis. Epidemiological studies in Chile [34, 35] have reported that exposure to high levels of iAs in infancy increased the risk of developing bladder and lung cancer in older age. Since the iAs exposure level in Asian countries had increased around 1980, a peak of carcinogenesis is expected around 2010–2030 if the incubation period for carcinogenesis is correct. Recently, reports of bladder cancer due to iAs exposure have increased [35, 48, 49]. On the other hand, some reports predicted a relatively low incidence of skin cancer in China [51].

Strangely, little progress has been made in elucidating the mechanism of arsenic carcinogenesis. Because the accumulation of knowledge that arsenic has none or weakly mutagenic property, researchers shifted their focus on the mechanisms involving epigenetic dysregulation in the carcinogenicity arsenic [52, 53].

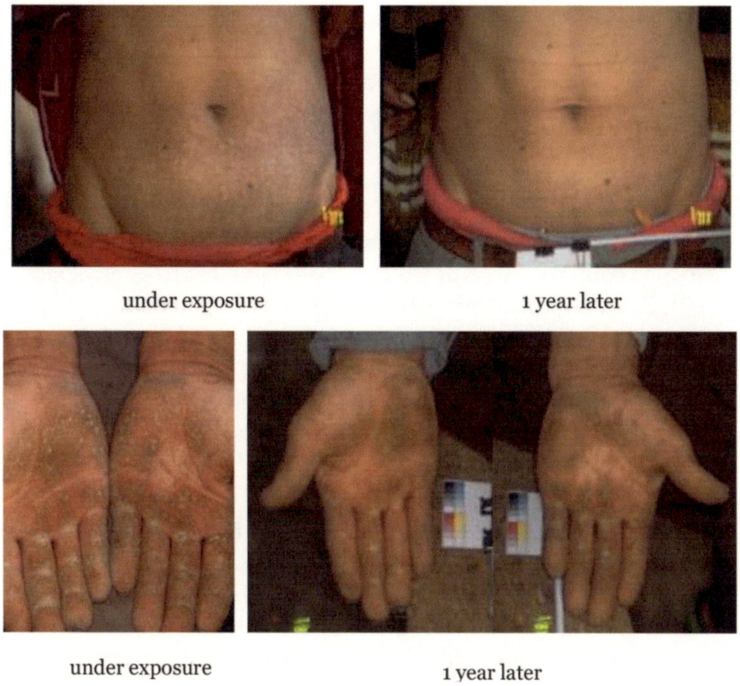

Figure 2.
Improvement of skin lesions by reducing iAs exposure for 1 year. Dyspigmentation at top, and hyperkeratosis at bottom.

4.3 iAs exposure and cognitive dysfunction

A growing number of studies on iAs exposure and health effects have focused on the cognitive abilities. A prospective study of subacute arsenic poisoning in infants led to find the cognitive dysfunction manifested in the elderly as a sequela of cerebral and central nervous system damage [4]. The outcomes are important as predictive data for cognitive dysfunction in chronic arsenic poisoning.

Studies in Bangladesh [54, 55], Taiwan [56], Mexico [57], Spain [58], and United States [59] confirmed cognitive abilities in children. Taiwanese studies reported that a lower AMC can lead to cognitive development [56]. The WHO (2018) raised the alarm about cognitive abilities from iAs exposure during fetal and infancy stages [60]. Recently, it has become known that rice accumulates arsenic in the soil at a high rate during cultivation. Since rice is a staple food in most Asian countries (which account for approximately 50% of the world's population), it is likely that the discussion of the sources of iAs exposure would revolve around rice, rather than drinking water. In this regard, EFSA discusses the need for the legal regulation of iAs intake from rice and processed rice products for infants and children [61].

Although the effects on cognitive abilities have been studied mainly in children, an association between iAs exposure and cognitive dysfunction in adults has been suggested [62–64]. However, there is insufficient information on the mechanism of cognitive dysfunction, whether it is due to the exposure in infancy which was manifested in older age [4] or to direct exposure in adults. iAs exposure started to be among the stated causes of cognitive dysfunction.

4.4 iAs exposure and cardiovascular disease (CVD) and diabetes

Exposure to high concentrations of iAs has been associated with CVD. The relationship between iAs exposure and CVD has been attributed to carotid atherosclerosis [65, 66], with an increase in %MMA, suggesting an association with a decrease in AMC. Nevertheless, no significant association between hypertension and a decreased AMC has been observed. We were the first to discover that peripheral vascular damage in iAs-exposed humans is induced by nitric oxide (NO), an oxidative stress [67].

Exposure to high concentrations of iAs has been suggested to be associated with diabetes [68, 69]. However, the relationship between AMC and diabetes at low iAs exposures has been studied, but no clear association has been found. The relationship between iAs intake (mainly from rice) and lifestyle-related diseases (non-communicable diseases, NCDs) is expected to be the subject of much debate among most people living in non-arsenic-contaminated environments.

5. Health effects of organic arsenic compounds ingested from seafood and seaweeds

The demand for seafood is increasing internationally as the dietary intake of omega-3 fatty acids, mainly found in seafood, has been highly recommended for the prevention of NCDs. On the other hand, arsenolipids, arsenobetaine, and arsenosugars are found in fish and shellfish, and varying toxicity has been observed (**Figure 3**). In 2022, the EFSA started to adopt legal regulations for the dietary intake of AsLipids in fish and shellfish [7].

5.1 Arsenolipids (AsLipids)

Many types of AsLipids have been detected in marine organisms, among which four were identified, including arsenic-containing hydrocarbons (AsHCs), arsenic-containing fatty acids (AsFAs), arsenic-containing phospholipids, and arsenic-containing phosphatidyl-cholines. Of these, the first two are noted in this study [5, 6, 70].

One of the major adverse health effects of AsLipids concerns brain function. An experimental validation suggests that AsHCs can penetrate the BBB and migrate to the brain tissues. An *in vitro* model of the blood–cerebrospinal fluid barrier composed of porcine choroid plexus epithelial cells showed that AsHCs were 1.5-fold more potent than arsenite (iAs^{III}) in disrupting the barrier function. Consequently, it is suggested that being exposed to high levels of certain AsLipids may be harmful to the central nervous system [71]. In addition, AsHCs have been shown to accumulate in brain tissues when applied to drosophila [72]. The toxicity of AsHCs in brain cells has been shown to be 5–19 times greater than that of iAs^{III} when applied to human neuronal astrocytes [73]. In contrast, AsFAs tend to be less permeable to the BBB and less neurotoxic than AsHCs [74].

To our knowledge, there is insufficient information to evaluate the effects of AsLipids on the human brain. After 24 h of salmon ingestion containing AsHCs, approximately 3% was detected in breast milk [75], which could cross the BBB and migrate to the infant's brain tissue. Alternatively, the strong brain cytotoxicity of AsHCs is also predicted to cause brain dysfunction.

Detailed studies on the biotransformation of AsLipids in animal studies are deemed difficult to conduct, due to the required large sample size. To search for AsLipid

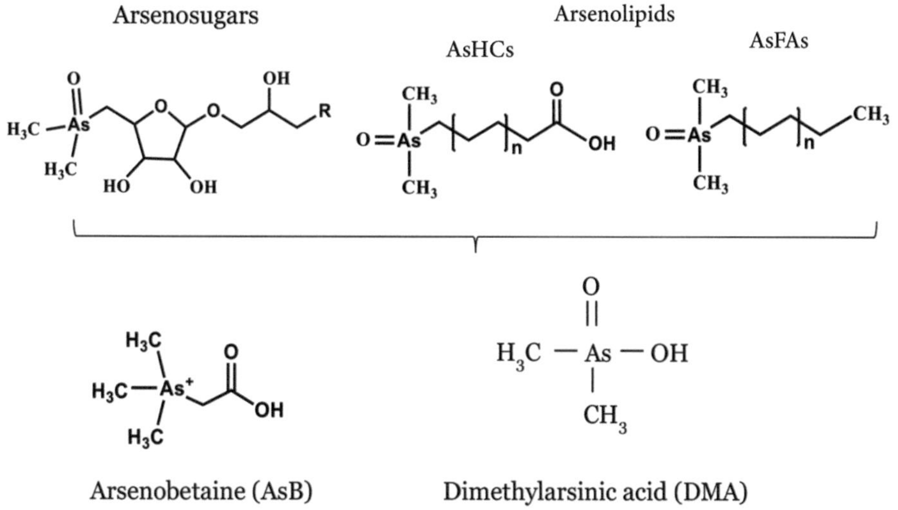

Figure 3.
Chemical structure of arsenolipids, arsenosugars, and arsenobetaine. Typical arsenolipids are known as arsenic-containing hydrocarbons (AsHCs) and arsenic-containing fatty acids (AsFAs). The major metabolite of arsenolipids and arsenosugars is DMA.

metabolites, the urinary arsenic was measured in two volunteers who took a single oral dose of test material prepared from cod liver oil. The major metabolites of AsLipids were DMA and small amounts of thio-dimethylarsenobutanoic acid (DMAB), oxo-DMAB, thio-dimethylarsenopropanoic acid (DMAPr), and oxo-DMAPr [76].

Studies predicting the daily intake of AsLipids are difficult and limited. However, a market basket survey estimated the daily intake of AsLipids from 152 food items, including seafood and algae. AsHCs and AsFAs were 4200 and 2100 ng As/person/day, respectively, as solely detected in seafood and algae [77]. Notably, AsLipid concentrations were measured in 18 samples of the purchased seafood, and AsHC values were 83 ± 73 ng As/g fresh-weight; this study revealed that consumed seafood contain toxic AsHC [78]. Seafood is commonly cooked by boiling or baking. The chemical structure of AsLipids, when heated in a microwave oven, was reported unchanged [79]. Therefore, toxic AsLipids are absorbed without loss due to cooking.

5.2 Arsenobetaine (AsB)

AsB in seawater is produced by the methylation of iAs by phytoplankton and zooplankton and is concentrated in fish or shellfish through the food chain. In mammals, AsB is not synthesized in the body; thus, it is entirely derived from seafood and is rapidly excreted in the urine, as verified in human [80] and animal [81, 82] studies. Its half lethal dose (LD_{50}) is 10 g/kg, and AsB has been shown to be the least toxic of the arsenic compounds [5, 6].

5.3 Arsenosugars (AsSugars)

The structure of AsSugars was first elucidated from brown kelp (*Ecklonia radiata*) by Edmonds and Francesconi in 1981 [83]. It is presumed that AsSugars in seaweeds are originally biosynthesized by the direct absorption of iAs from seawater. Seaweed and kelp showed to contain large amounts of AsSugar-Gly and AsSugar-SO_3, which are not degraded into iAs when treated with heat or stomach acid [6]. The metabolites of AsSugars, validated in humans, were DMA, thio-dimethylarsinoylacetate, thio-dimethylarsinoylethanol, and thio-dimethyl arsenate [84]. The last two seem to be the most toxic, and their cytotoxicity is suggested to be similar to that of iAs^{III} [85]. The major metabolite of AsSugars in human urine is DMA, which is less toxic than iAs. Further monitoring is needed due to the limited knowledge of the metabolism and toxicity of AsSugars.

6. Abandoned arsenic chemical weapons and health effects

Chemical weapons (CW) were produced in large amounts by the countries that participated in World Wars I and II. In 1997, the Chemical Weapons Convention was signed by the international community, which now includes 192 countries. The Organization for the Prohibition of Chemical Weapons (OPCW, Hague, the Netherlands) oversees and operates the disposal of CW [86]. OPCW defines CW as any chemical substance used to intentionally cause death or harm by its toxicity. The most potent chemical warfare agents include the blister agents: sulfur or nitrogen mustards (yperite), and lewisite; the riot control (vomiting) agents: diphenylchloroarsine (DA, Clark I), diphenylcyanoarsine (DC, Clark II), and adamsite; the blood agents: hydrogen cyanide and arsine; the asphyxiants: chlorine and phosgene; and the nerve agents: sarin, soman, tabun, VX, and Novichok. The main arsenic CW (**Figure 4**) produced in World War II were the blister and the vomiting agents, which have been largely produced in Japan, Germany, the Soviet Union, and the United States [87].

There are two types of CW: stockpile (available) and non-stockpile (abandoned). So far, the largest stockpiles have been held by the United States and Russia, both of which have approximately 30,000–40,000 tons. The United States declared that most of its stockpiled CW have been completely destroyed [87]; however, information

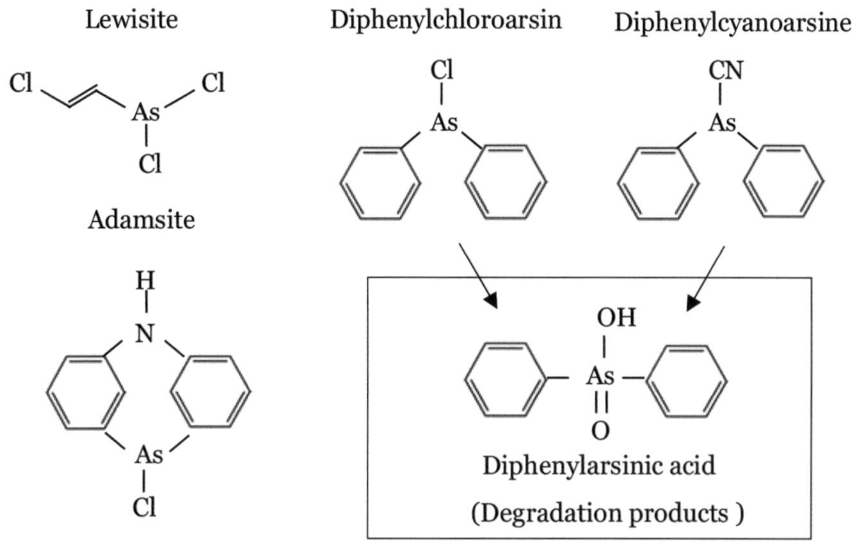

Figure 4.
Chemical structure of arsenic chemical weapons. The degradation products of diphenylchloroarsin (DA, Clark I) and diphenylcyanoarsine (DC, Clark II) are diphenylarsinic acid (DPAA).

about the destruction process in Russia is still not credible. The international community has major concerns related to the non-signatory countries that still possess CW, in addition to the fears of CW possession and use by terrorist organizations. After World War II, many countries dumped their CW in the sea as a cheap disposal method [87]. Approximately 50,000 tons of arsenic CW (including Clarks I and II and adamsite) are estimated to be left in the Baltic Sea [8], which has resulted in phenylarsenic chemical contamination of some seafood [8–10]. The former Japanese military buried large quantities of arsenic CW in various parts of China during its defeat in World War II. In particular, iAs and phenylarsenic chemical warfare agents derived from arsenic CW have been recently detected in agricultural soil in Jilin Province [12–14]. Alternatively, the arsenic contamination persists in Germany at the sites of the facilities where arsenic CW filling operations were conducted [11]. After more than 75 years of arsenic CW abandonment, the derived risk of health effects from arsenic still exists in soils and the seabed. Therefore, the international community is invited to recognize the current situation and take action accordingly.

6.1 Lewisite

The blister agents include mustard and lewisite. The former freezes in cold climates ($\leq -20°C$) and loses its function as a weapon, while the latter, to which arsenic is added, does not freeze and is used in cold climates. Therefore, lewisite was manufactured in larger amounts by the former Japanese Army and the Soviet Union than in any other country. In 2003, 44 people at a construction site in Heilongjiang Province in China, suffered from severe skin damage due to exposure to containers filled with arsenical CWs (lewisite + mustard), believed to have been abandoned by the former Japanese army [88]. Consequently, this incident implies that the arsenic-containing CW, lewisite, has not been completely and safely removed from the environment.

Exposure to lewisite can cause rapid hydrolysis in the epidermis, resulting in severe skin blistering and inflammation, after which the arsenic is absorbed through the dermis [89]. The eyes are sensitive to lewisite exposure, which can cause edema, blepharospasm, and blindness [90]. Lewisite is absorbed transdermally and distributed throughout the body, causing multiple organ damage in the lungs [91] and kidneys [92], and increasing the likelihood of mortality. On the other hand, little is known about the sequelae following lewisite and mustard exposure. A > 10-year follow-up study on the 44 victims exposed at construction work sites revealed mental and cognitive dysfunctions as new arising health problems [88].

6.2 Diphenylchloroarsine (DA) and diphenylcyanoarsine (DC)

Arsenic CWs that are known as vomiting agents, include DA, DC, and adamsite (**Figure 4**) and were aimed to be used for disturbing combat abilities of the soldiers rather than for killing or wounding them. Exposure to DA or DC can cause a strong sneeze, nausea, and instant vomiting. However, their detailed toxic effect is still unclear. Diphenylarsinic acid (DPAA) is the raw material for the production of DA and DC and, conversely, their degrading substance in the human body. In 2003, poisoning from contaminated well water in Kamisu, Japan, was caused by DPAA that had been buried underground. The residents used this water for cooking and bathing, while some others used it to prepare infant formula. Ishii et al. (2004) reported that in 2000, the first victim was a junior high school student who had cerebellar ataxia, tremors, and myoclonus [15]. Subsequently, a population health survey led by medical professionals revealed the presence of a total of 157 victims in the surrounding area. Before this incident, there was no toxicological data on DPAA. DPAA can pass through the BBB in trace amounts and target the cerebellum, causing disruption of the glutamatergic system [93, 94]. A summary document published by the Japanese Ministry of the Environment [95] revealed that DPAA is mainly absorbed in the gastrointestinal tract, while a mere amount is absorbed through the skin. Additionally, DPAA is not metabolized in the body and is mainly excreted in the urine. The relative toxicity of DPAA deduced from cytotoxicity studies is 1/100th compared with ATO.

So far, information on DA and DC has been limited to their acute toxicity as vomiting agents, damaging the mucous membranes of the eyes and nasal passages. We did not envision a situation where DA and DC would be decomposed to product DPAA, which would then be ingested *via* drinking water. However, DPAA and related substances have been found to contaminate fish, shellfish, and soil [8–14]. The clinical findings in DPAA poisoning patients [15] and animal studies [93–95] are consistent with the brain being the target organ. Toxicological information on DPAA suggests that it may be the most dangerous of the arsenic compounds. In the future, the factual extent of contamination in seafood, agricultural products, and drinking water needs to be further investigated.

7. Conclusions

The AMC index determined from non-confounding urinary arsenic in patients with acute arsenic poisoning is expected to contribute to research investigations of chronic arsenic poisoning. The effects of acute and chronic arsenic poisoning on health are associated with decreased AMC. Children and adult females tend to have higher levels of AMC than adult males. However, iAs exposure in infancy increases cancer and cognitive dysfunction in old age. Furthermore, AsLipids ingested from seafood penetrate the BBB and accumulate in the brain, resulting in toxicity. DPAA, a degradant of DA and DC used in vomiting agents among the arsenic CWs used in World War II, is overlooked by the international community as a cause of cerebellar disorders. In the past, the effects of arsenic on health have been mostly concerned with carcinogenesis. However, the target organ for iAs, AsLipids, and DPAA is mainly the brain; therefore, a deeper understanding of brain dysfunction is needed. Scientific evidence has begun to confirm that exposure to iAs during infancy and conception (fetus) exacerbates carcinogenesis and cognitive dysfunction in old age.

Acknowledgements

This work was supported by JSPS KAKENHI Grant Number JP17H04655 and JP21H03185.

Conflict of interest

The authors declare no conflict of interest.

Author details

Hiroshi Yamauchi[1*], Takahiko Yoshida[2] and Ayako Takata[1]

1 Department of Preventive Medicine, St. Marianna University School of Medicine, Kawasaki, Japan

2 Department of Social Medicine, Asahikawa Medical University, Asahikawa, Japan

*Address all correspondence to: hyama@marianna-u.ac.jp

References

[1] Vahter ME. Interactions between arsenic-induced toxicity and nutrition in early life. Journal of Nutrition. 2007;**137**(12):2798-2804. DOI: 10.1093/jn/137.12.2798

[2] Shen H, Niu O, Xu M, Rui D, Xu S, Feng G, et al. Factors affecting arsenic methylation in arsenic-exposed humans: A systematic review and meta-analysis. International Journal of Environmental Research and Public Health. 2016;**13**(2):205. DOI: 10.3390/ijerph13020205

[3] Yamauchi H, Takata A. Arsenic metabolism differs between child and adult patients during acute arsenic poisoning. Toxicology and Applied Pharmacology. 2021;**410**:115352. DOI: 10.1016/j.taap.2020.115352

[4] Liu R, Tabuchi T, Kitamura T, Miyashiro I, Sobue T. Long-term observational study on 6223 survivors of arsenic poisoning due to contaminated milk powder during infancy. Cancer Science. 2020;**111**(10):3873-3880. DOI: 10.1111/cas.14623

[5] Taylor V, Goodale B, Raab A, Schwerdtle T, Reimer K, Conklin S, et al. Human exposure to organic arsenic species from seafood. Science of the Total Environment. 2017;**580**:266-282. DOI: 10.1016/j.scitotenv.2016.12.113

[6] Cao Y, Takata A, Hitomi T, Yamauchi H. Metabolism and toxicity of organic arsenic compounds in marine organisms. In: Yamauchi H, Sun G, editors. Arsenic Contamination in Asia. Current Topics in Environmental Health and Preventive Medicine. Singapore: Springer; 2018. pp. 119-136. DOI: 10.1007/978-981-13-2565-6_7

[7] European Food Safety Authority (EFSA). Extensive literature search on organic arsenic in food. EFSA Journal. 2022;**19**(9):EN-7565. DOI: 10.2903/sp.efsa.2022.EN-7565

[8] Vanninen P, Östin A, Bełdowski J, Pedersen EA, Söderström M, Szubska M, et al. Exposure status of sea-dumped chemical warfare agents in the Baltic Sea. Marine Environmental Research. 2020;**161**:105112. DOI: 10.1016/j.marenvres.2020.105112

[9] Niemikoski H, Lehtonen KK, Ahvo A, Heiskanen I, Vanninen P. Metabolism and cytotoxicity of diphenylarsinic acid, a degradation product of sea-dumped chemical warfare agents, in a rainbow trout liver cell line RTL-W1. Aquatic Toxicology. 2021;**241**:105993. DOI: 10.1016/j.aquatox.2021.105993

[10] Czub M, Nawała J, Popiel S, Brzeziński T, Maszczyk P, Sanderson H, et al. Acute aquatic toxicity of arsenic-based chemical warfare agents to Daphnia magna. Aquatic Toxicology. 2021;**230**:105693. DOI: 10.1016/j.aquatox.2020.105693

[11] Daus B, Hempel M, Wennrich R, Weiss H. Concentrations and speciation of arsenic in groundwater polluted by warfare agents. Environmental Pollution. 2010;**158**:3439-3444. DOI: 10.1016/j.envpol.2010.06.029

[12] Zhang Y, Xi H, Zhu Y, Zhao S, Ji C. Study on the composition and distribution characteristics of As in As-containing agent contaminated soil. Environmental Research. 2022;**214**(Pt 3):114039. DOI: 10.1016/j.envres.2022.114039

[13] Zhang Y, Zhu Y, Zhao S, Li D, Xi H, Wang Y. Arsenic in soils contaminated by arsenic-containing chemical weapons in a site of Jilin, China: Fraction

and bioaccessibility. Environmental Science and Pollution Research. 2022;**29**(19):28957-28972. DOI: 10.1007/s11356-021-18482-3

[14] Ji C, Zhu Y, Zhao S, Zhang H, Wang S, Zhou J, et al. Arsenic and heavy metals at Japanese abandoned chemical weapons site in China: Distribution characterization, source identification and contamination risk assessment. Environmental Geochemistry and Health. 2022. DOI: 10.1007/s10653-022-01382-x

[15] Ishii K, Tamaoka A, Otsuka F, Iwasaki N, Shin K, Matsui A, et al. Diphenylarsinic acid poisoning from chemical weapons in Kamisu, Japan. Annals of Neurology. 2004;**56**(5):741-745. DOI: 10.1002/ana.20290

[16] Jolliffe DM, Budd AJ, Gwilt DJ. Massive acute arsenic poisoning. Anaesthesia. 1991;**46**(4):288-290. DOI: 10.1111/j.1365-2044.1991.tb11500.x

[17] Kamijo Y, Soma K, Asari Y, Ohwada T. Survival after massive arsenic poisoning self-treated by high fluid intake. Journal of Toxicology. Clinical Toxicology. 1998;**36**(1-2):27-29. DOI: 10.3109/15563659809162579

[18] Heinrich-Ramm R, Schaller H, Horn J, Angerer J. Arsenic species excretion after dimercaptopropanesulfonic acid (DMPS) treatment of an acute arsenic trioxide poisoning. Archives of Toxicology. 2003;**77**:63-68. DOI: 10.1007/s00204-002-0413-z

[19] Dueñas-Laita A, Pérez-Miranda M, González-López MA, Martín-Escudero JC, Ruiz Mambrilla M, Blanco-Varela J. Acute arsenic poisoning. Lancet. 2005;**365**(9475):1982. DOI: 10.1016/S0140-6736(05)66670-6

[20] Lu PH, Tseng JC, Chen CK, Chen CH. Survival without peripheral neuropathy after massive acute arsenic poisoning: Treated by 2,3-dimercaptopropane-1-sulphonate. Journal of Clinical Pharmacy and Therapeutics. 2017;**42**(4):506-508. DOI: 10.1111/jcpt.12538

[21] Vallee BL, Ulmer DD, Wacker WEC. Arsenic toxicology and biochemistry. Archives of Industrial Health. 1960;**21**:132-151

[22] Uede K, Furukawa F. Skin manifestations in acute arsenic poisoning from the Wakayama curry-poisoning incident. British Journal of Dermatology. 2003;**149**:757-762. DOI: 10.1046/j.1365-2133.2003.05511.x

[23] Kishi Y, Sasaki H, Yamasaki H, Ogawa K, Nishi M, Nanjo K. An epidemic of arsenic neuropathy from a spiked curry. Neurology. 2001;**56**(10):1417-1418. DOI: 10.1212/wnl.56.10.1417

[24] Nurchi VN, Djordjevic AB, Crisponi G, Alexander J. Arsenic toxicity: Molecular targets and therapeutic agents. Biomolecules. 2020;**10**(2):235. DOI: 10.3390/biom10020235

[25] Bjørklund G, Oliinyk P, Lysiuk R, Rahaman MS, Antonyak H, Lozynska I, et al. Arsenic intoxication: General aspects and chelating agents. Archives of Toxicology. 2020;**94**(6):1879-1897. DOI: 10.1007/s00204-020-02739-w

[26] Waalkes MP, Liu J. Early-life arsenic exposure: Methylation capacity and beyond. Environmental Health Perspectives. 2008;**116**(3):A104. DOI: 10.1289/ehp.11276

[27] Lindberg AL, Ekström EC, Nermell B, Rahman M, Lönnerdal B, Persson LA, et al. Genderand age differences in the metabolism of inorganic arsenic in a highly exposed population in Bangladesh. Environmental Research. 2008;**106**(1):110-120. DOI: 10.1016/j.envres.2007.08.011

[28] Skröder Löveborn H, Kippler M, Lu Y, Ahmed S, Kuehnelt D, Raqib R, et al. Arsenic metabolism in children differs from that in adults. Toxicological Sciences. 2016;**152**(1):29-39. DOI: 10.1093/toxsci/kfw060

[29] Vahter M. Effects of arsenic on maternal and fetal health. Annual Review of Nutrition. 2009;**29**:381-399. DOI: 10.1146/annurev-nutr-080508-141102

[30] Desai G, Millen AE, Vahter M, Queirolo EI, Peregalli F, Mañay N, et al. Associations of dietary intakes and serum levels of folate and vitamin B-12 with methylation of inorganic arsenic in Uruguayan children: Comparison of findings and implications for future research. Environmental Research. 2020;**189**:109935. DOI: 10.1016/j.envres.2020.109935

[31] Abuawad A, Bozack AK, Saxena R, Gamble MV. Nutrition, one-carbon metabolism and arsenic methylation. Toxicology. 2021;**457**:152803. DOI: 10.1016/j.tox.2021.152803

[32] Yamauchi H, Takata A. Past and current arsenic poisonings. In: Yamauchi H, Sun G, editors. Arsenic Contamination in Asia, Biological Effects and Preventive Measures. Singapore: Springer; 2018. pp. 1-11. DOI: 10.1007/978-981-13-2565-6_1

[33] Tanaka H, Tsukuma H, Oshima A. Long-term prospective study of 6104 survivors of arsenic poisoning during infancy due to contaminated milk powder in 1955. Journal of Epidemiology. 2010;**20**(6):439-445. DOI: 10.2188/jea.JE20090131

[34] Roh T, Steinmaus C, Marshall G, Ferreccio C, Liaw J, Smith AH. Age at exposure to arsenic in water and mortality 30-40 years after exposure cessation. American Journal of Epidemiology. 2018;**187**(11):2297-2305. DOI: 10.1093/aje/kwy159

[35] Smith AH, Marshall G, Roh T, Ferreccio C, Liaw J, Steinmaus C. Lung, bladder, and kidney cancer mortality 40 years after arsenic exposure reduction. Journal of the National Cancer Institute. 2018;**110**(3):241-249. DOI: 10.1093/jnci/djx201

[36] Yoshida T, Yamauchi H, Fan SG. Chronic health effects in people exposed to arsenic via the drinking water: Dose-response relationships in review. Toxicology and Applied Pharmacology. 2004;**198**(3):243-252. DOI: 10.1016/j.taap.2003.10.022

[37] De Loma J, Skröder H, Raqib R, Vahter M, Broberg K. Arsenite methyltransferase (AS3MT) polymorphisms and arsenic methylation in children in rural Bangladesh. Toxicology and Applied Pharmacology. 2018;**357**:80-87. DOI: 10.1016/j.taap.2018.08.020

[38] De Loma J, Vicente M, Tirado N, Ascui F, Vahter M, Gardon J, et al. Human adaptation to arsenic in Bolivians living in the Andes. Chemosphere. 2022;**301**:134764. DOI: 10.1016/j.chemosphere.2022.134764

[39] Yamauchi H, Yamamura Y. Metabolism and excretion of orally administered dimethylarsinic acid in the hamster. Toxicology and Applied Pharmacology. 1984;**74**(1):134-140. DOI: 10.1016/0041-008x(84)90279-5

[40] Tamayo LI, Kumarasinghe Y, Tong L, Balac O, Ahsan H, Gamble M, et al. Inherited geneticneffects on arsenic metabolism: A comparison of effects on arsenic species measured in urine and in blood. Environmental Epidemiology. 2022;**6**(6):e230. DOI: 10.1097/EE9.0000000000000230

[41] Yoshida T, Sun G, Pi J, Li X, Li B, Yamauchi H. Field research on chronical arsenic poisoning in Inner Mongolia, China. In: Yamauchi H, Sun G, editors. Arsenic Contamination in Asia, Biological Effects and Preventive Measures. Singapore: Springer; 2018. pp. 61-81. DOI: 10.1007/978-981-13-2565-6_5

[42] Seow WJ, Pan WC, Kile ML, Baccarelli AA, Quamruzzaman Q, Rahman M, et al. Arsenic reduction in drinking water and improvement in skin lesions: A follow-up study in Bangladesh. Environmental Health Perspectives. 2012;**120**(12):1733-1738. DOI: 10.1289/ehp.1205381

[43] Wei B, Yu J, Kong C, Li H, Yang L, Xia Y, et al. Effects of arsenic methylation and metabolism on the changes of arsenic-related skin lesions. Environmental Science and Pollution Research. 2018;**25**(24):24394-24402. DOI: 10.1007/s11356-018-2512-2

[44] IARC, Arsenic, metals, fibres, and dusts. IARC Monographs on the Evaluation of Carcinogenic Risks to Humans. 100C. Lyon. 2012

[45] Hsueh YM, Chiou HY, Huang YL, Wu WL, Huang CC, Yang MH, et al. Serum beta-carotene level, arsenic methylation capability, and incidence of skin cancer. Cancer Epidemiology Biomarkers and Prevention. 1997;**6**(8):589-596

[46] Yu RC, Hsu KH, Chen CJ, Froines JR. Arsenic methylation capacity and skin cancer. Cancer Epidemiology Biomarkers and Prevention. 2000;**9**(11):1259-1262

[47] Chen YC, Guo YL, Su HJ, Hsueh YM, Smith TJ, Ryan LM, et al. Arsenic methylation and skin cancer risk in southwestern Taiwan. Journal of Occupational and Environmental Medicine. 2003;**45**(3):241-248. DOI: 10.1097/01.jom.0000058336.05741.e8

[48] Huang YK, Huang YL, Hsueh YM, Yang MH, Wu MM, Chen SY, et al. Arsenic exposure, urinary arsenic speciation, and the incidence of urothelial carcinoma: A twelve-year follow-up study. Cancer Causes & Control. 2008;**19**(8):829-839. DOI: 10.1007/s10552-008-9146-5

[49] Melak D, Ferreccio C, Kalman D, Parra R, Acevedo J, Pérez L, et al. Arsenic methylation and lung and bladder cancer in a case-control study in northern Chile. Toxicology and Applied Pharmacology. 2014;**274**(2):225-231. DOI: 10.1016/j.taap.2013.11.014

[50] Steinmaus C, Yuan Y, Kalman D, Rey OA, Skibola CF, Dauphine D, et al. Individual differences in arsenic metabolism and lung cancer in a case-control study in Cordoba. Argentina. Toxicology and Applied Pharmacology. 2010;**247**(2):138-145. DOI: 10.1016/j.taap.2010.06.006

[51] Huang L, Wu H, van der Kuijp TJ. The health effects of exposure to arsenic-contaminated drinkingwater: A review by global geographical distribution. International journal of environmental health research. 2015;**25**(4):432-452. DOI: 10.1080/09603123.2014.958139

[52] Saintilnord WN, Fondufe-Mittendorf Y. Arsenic-induced epigenetic changes in cancer development. Seminars in Cancer Biology. 2021;**76**:195-205. DOI: 10.1016/j.semcancer.2021.03.019

[53] Islam R, Zhao L, Wang Y, Lu-Yao G, Liu LZ. Epigenetic dysregulations in arsenic-induce carcinogenesis. Cancers. 2022;**14**(18):4502. DOI: 10.3390/cancers14184502

[54] Vahter M, Skröder H, Rahman SM, Levi M, Derakhshani Hamadani J,

Kippler M. Prenatal and childhood arsenic exposure through drinking water and food and cognitive abilities at 10 years of age: A prospective cohort study. Environment International. 2020;**139**:105723. DOI: 10.1016/j.envint.2020.105723

[55] Hamadani JD, Tofail F, Nermell B, Gardner R, Shiraji S, Bottai M, et al. Critical windows of exposure for arsenic-associated impairment of cognitive function in pre-school girls and boys: A population-based cohort study. International Journal of Epidemiology. 2011;**40**(6):1593-1604. DOI: 10.1093/ije/dyr176

[56] Hsueh YM, Lin YC, Chung CJ, Huang YL, Hsieh RL, Huang PT, et al. Combined effect of polymorphisms of MTHFR and MTR and arsenic methylation capacity on developmental delay in preschool children in Taiwan. Archives of Toxicology. 2020;**94**(6):2027-2038. DOI: 10.1007/s00204-020-02745-y

[57] Rosado JL, Ronquillo D, Kordas K, Rojas O, Alatorre J, Lopez P, et al. Arsenic exposure and cognitive performance in Mexican schoolchildren. Environmental Health Perspectives. 2007;**115**(9):1371-1375. DOI: 10.1289/ehp.9961

[58] Soler-Blasco R, Murcia M, Lozano M, Sarzo B, Esplugues A, Riutort-Mayol G, et al. Prenatal arsenic exposure, arsenic methylation efficiency, and neuropsychological development among preschool children in a Spanish birth cohort. Environmental Research. 2022;**207**:112208. DOI: 10.1016/j.envres.2021.112208

[59] Signes-Pastor AJ, Romano ME, Jackson B, Braun JM, Yolton K, Chen A, et al. Associations of maternal urinary arsenic concentrations during pregnancy with childhood cognitive abilities: The HOME study. International Journal of Hygiene and Environmental Health. 2022;**245**:114009. DOI: 10.1016/j.ijheh.2022.114009

[60] WHO. Arsenic. Available from: https://www.who.int/news-room/fact-sheets/detail/arsenic [Accessed: March 8, 2023]

[61] European Food Safety Authority (EFSA); Arcella D, Cascio C, Gómez Ruiz JÁ. Chronic dietary exposure to inorganic arsenic. EFSA Journal. 2021;**19**(1):e06380. DOI: 10.2903/j.efsa.2021.-6380

[62] Liu J, Gao Y, Liu H, Sun J, Liu Y, Wu J, et al. Assessment of relationship on excess arsenic intake from drinking water and cognitive impairment in adults and elders in arsenicosis areas. International Journal of Hygiene and Environmental Health. 2017;**220**(2 Pt B):424-430. DOI: 10.1016/j.ijheh.2016

[63] Karim Y, Siddique AE, Hossen F, Rahman M, Mondal V, Banna HU, et al. Dose-dependent relationships between chronic arsenic exposure and cognitive impairment and serum brain-derived neurotrophic factor. Environment International. 2019;**131**:105029. DOI: 10.1016/j.envint.2019.105029

[64] Wang X, Huang X, Zhou L, Chen J, Zhang X, Xu K, et al. Association of arsenic exposure and cognitive impairment: A population-based cross-sectional study in China. Neurotoxicology. 2021;**82**:100-107. DOI: 10.1016/j.neuro.2020.11.009

[65] Chen Y, Wu F, Liu M, Parvez F, Slavkovich V, Eunus M, et al. A prospective study of arsenic exposure, arsenic methylation capacity, and risk of cardiovascular disease in Bangladesh. Environmental Health Perspectives. 2013;**121**(7):832-838. DOI: 10.1289/ehp.1205797

[66] Moon KA, Zhang Y, Guallar E, Francesconi KA, Goessler W, Umans JG, et al. Association of low-moderate urine arsenic and QT interval: Cross-sectional and longitudinal evidence from the strong heart study. Environmental pollution. 2018;**240**:894-902. DOI: 10.1016/j.envpol.2018.04.129

[67] Pi J, Yamauchi H, Sun G, Yoshida T, Aikawa H, Fujimoto W, et al. Vascular dysfunction in patients with chronic arsenosis can be reversed by reduction of arsenic exposure. Environmental Health Perspectives. 2005;**113**(3):339-341. DOI: 10.1289/ehp.7471

[68] Kuo CC, Moon KA, Wang SL, Silbergeld E, Navas-Acien A. The association of arsenic metabolism with cancer, cardiovascular disease, and diabetes: A systematic review of the epidemiological evidence. Environmental Health Perspectives. 2017;**125**(8):087001. DOI: 10.1289/EHP577

[69] Farkhondeh T, Samarghandian S, Azimi-Nezhad M. The role of arsenic in obesity and diabetes. Journal of Cellular Physiology. 2019;**234**(8):12516-12529. DOI: 10.1002/jcp.28112

[70] Luvonga C, Rimmer CA, Yu LL, Lee SB. Organoarsenicals in seafood: Occurrence, dietary exposure, toxicity, and risk assessment considerations - a review. Journal of Agricultural and Food Chemistry. 2020;**68**(4):943-960. DOI: 10.1021/acs.jafc.9b07532

[71] Müller SM, Ebert F, Bornhorst J, Galla HJ, Francesconi KA, Schwerdtle T. Arsenic-containing hydrocarbons disrupt a model in vitro blood-cerebrospinal fluid barrier. Journal of Trace Elements in Medicine and Biology. 2018;**49**:171-177. DOI: 10.1016/j.jtemb.2018.01.020

[72] Niehoff AC, Schulz J, Soltwisch J, Meyer S, Kettling H, Sperling M, et al. Imaging by elemental and molecular mass spectrometry reveals the uptake of an arsenolipid in the brain of drosophila melanogaster. Analytical Chemistry. 2016;**88**(10):5258-5263. DOI: 10.1021/acs.analchem.6b00333

[73] Witt B, Bornhorst J, Mitze H, Ebert F, Meyer S, Francesconi KA, et al. Arsenolipids exert less toxicity in a human neuron astrocyte co-culture as compared to the respective monocultures. Metallomics. 2017;**9**(5):442-446. DOI: 10.1039/c7mt00036g

[74] Meyer S, Raber G, Ebert F, Leffers L, Müller SM, Taleshi MS, et al. In vitro toxicological characterisation of arsenic-containing fatty acids and three of their metabolites. Toxicology Research. 2015;**4**(5):1289-1296. DOI: 10.1039/c5tx00122f

[75] Xiong C, Stiboller M, Glabonjat RA, Rieger J, Paton L, Francesconi KA. Transport of arsenolipids to the milk of a nursing mother after consuming salmon fish. Journal of Trace Elements in Medicine and Biology. 2020;**61**:126502. DOI: 10.1016/j.jtemb.2020.126502

[76] Amayo KO, Raab A, Krupp EM, Feldmann J. Identification of arsenolipids and their degradation products in cod-liver oil. Talanta. 2014;**118**:217-223. DOI: 10.1016/j.talanta.2013.-09.056

[77] Amin MHA, Xiong C, Glabonjat RA, Francesconi KA, Oguri T, Yoshinaga J. Estimation of daily intake of arsenolipids in Japan based on a market basket survey. Food and Chemical Toxicology. 2018;**118**:245-251. DOI: 10.1016/j.fct.2018.05.019

[78] Amin MH, Xiong C, Francesconi KA, Itahashi Y, Yoneda M, Yoshinaga J. Variation in arsenolipid concentrations in seafood consumed in Japan.

Chemosphere. 2020;**239**:124781. DOI: 10.1016/j.chemosphere.2019.124781

[79] Xiong C, Glabonjat RA, Al Amin MH, Stiboller M, Yoshinaga J, Francesconi KA. Arsenolipids in salmon are partly converted to thioxo analogs during cooking. Journal of Trace Elements in Medicine and Biology. 2022;**69**:126892. DOI: 10.1016/j.jtemb.2021.126892

[80] Yamauchi H, Yamamura Y. Metabolism and excretion of orally ingested trimethylarsenic in man. Bulletin of Environmental Contamination and Toxicology. 1984;**32**(6):682-687. DOI: 10.1007/BF01607556

[81] Vahter M, Marafante E, Dencker L. Metabolism of arsenobetaine in mice, rats and rabbits. Science of the Total Environment. 1983;**30**:197-211. DOI: 10.1016/0048-9697(83)90012-8

[82] Yamauchi H, Kaise T, Yamamura Y. Metabolism and excretion of orally administered in the hamster. Bulletin of Environmental Contamination and Toxicology. 1984;**36**(3):350-355. DOI: 10.1007/BF01623519

[83] Edmonds JS, Francesconi KA. Arseno-sugars from brown kelp (*Ecklonia radiata*) as intermediates in cycling of arsenic in a marine. Nature. 1981;**289**:602-604. DOI: 10.1038/289602a0

[84] Francesconi KA, Tanggaar R, McKenzie CJ, Goessler W. Arsenic metabolites in human urine after ingestion of an arsenosugar. Clinical Chemistry. 2002;**48**(1):92-101. DOI: 10.1093/clinchem/48.1.92

[85] Leffers L, Ebert F, Taleshi MS, Francesconi KA, Schwerdtle T. In vitro toxicological characterization of two arsenosugars and their metabolites. Molecular Nutrition & Food Research. 2013;**57**(7):1270-1282. DOI: 10.1002/mnfr.201200821

[86] Organisation for the Prohibition of Chemical Weapons (OPCW). Available from: https://www.opcw.org/about-us [Accessed: March 8, 2023]

[87] Muzaffar S, Khan J, Srivastava R, Gorbatyuk MS, Athar M. Mechanistic understanding of the toxic effects of arsenic and warfare arsenicals on human health and environment. Cell Biology and Toxicology. 2023;**39**(1):85-110. DOI: 10.1007/s10565-022-09710-8

[88] Isono O, Kituda A, Fujii M, Yoshinaka T, Nakagawa G, Suzuki Y. Long-term neurological and neuropsychological complications of sulfur mustard and lewisite mixture poisoning in Chinese victims exposed to chemical warfare agents abandoned at the end of WWII. Toxicology Letters. 2018;**293**:9-15. DOI: 10.1016/j.toxlet.2018.04.017

[89] Li C, Srivastava RK, Athar M. Biological and environmental hazards associated with exposure to chemical warfare agents: Arsenicals. Annals of the New York Academy of Sciences. 2016;**1378**(1):143-157. DOI: 10.1111/nyas.13214

[90] Tewari-Singh N, Croutch CR, Tuttle R, Goswami DG, Kant R, Peters E, et al. Clinical progression of ocular injury following arsenical vesicant lewisite exposure. Cutaneous and Ocular Toxicology. 2016;**35**(4):319-328. DOI: 10.3109/15569527.2015.1127255

[91] Manzoor S, Mariappan N, Zafar I, Wei CC, Ahmad A, Surolia R, et al. Cutaneous lewisite exposure causes acute lung injury. Annals of the New York Academy of Sciences. 2020;**1479**(1):210-222. DOI: 10.1111/nyas.14346

[92] Srivastava RK, Traylor AM, Li C, Feng W, Guo L, Antony VB, et al.

Cutaneous exposure to lewisite causes acute kidney injury by invoking DNA damage and autophagic response. American Journal of Physiology. 2018;**314**(6):F1166-F1176. DOI: 10.1152/ajprenal.-00277.2017

[93] Negishi T, Matsunaga Y, Kobayashi Y, Hirano S, Tashiro T. Developmental subchronic exposure to diphenylarsinic acid induced increased exploratory behavior, impaired learning behavior, and decreased cerebellar glutathione concentration in rats. Toxicological Sciences. 2013;**136**(2):478-486. DOI: 10.1093/toxsci/kft200

[94] Sasaki S, Negishi T, Tsuzuki T, Yukawa K. Diphenylarsinic acid induced activation of MAP kinases, transcription factors, and oxidative stress-responsive factors and hypersecretion of cytokines in cultured normal human cerebellar astrocytes. Neurotoxicology. 2022;**88**:196-207. DOI: 10.1016/j.neuro.2021.12.002

[95] Ministry of the Environment, Japan. Health Effects from diphenylarsinic Acid (Japanese report, translated by author). 2008. pp. 1-20

Chapter 5

Arsenic Toxicity in Fish: Sources and Impacts

Ayesha Malik, Fakhira Khalid, Nigah Hidait, Khalid Mehmood Anjum, Saima Mahad, Abdul Razaq, Hamda Azmat and Muhammad Bilal Bin Majeed

Abstract

Arsenic has become a major toxicological concern due to its rising concentrations in aquatic bodies. It is added to the water either by natural sources including weathering of rocks, sediments, volcanic eruptions and aquifers, or by anthropogenic sources including herbicides, wood preservatives, metal smelting, drugs, pesticides, burning of coal, agriculture runoff and petroleum refining processes among others. The untreated and uncontrolled discharge of arsenic by industries into the natural water bodies poses serious threat to aquatic fauna by deteriorating water quality and making it unsuitable for fishes. Fish is an important bioindicator of aquatic bodies and excessive arsenic concentration causes its bioaccumulation in fish organs and muscles. This deposited arsenic in the fish imposes serious damage to physiology, biochemical disorders such as poisoning of gills, livers, decrease fertility, tissue damage, lesions, and cell death. It also enters in the cell and produces reactive oxygen species which increases the level of stress which further concentrates the oxidative enzymes and cortisol levels in fish. The uncontrolled discharge of arsenic and its devastating impact on fish diversity is a major concern for aquaculture progress and economic stability. This, along with its other implications is the scope of this chapter.

Keywords: arsenic toxicity, fish, arsenic sources, aquatic arsenic, arsenic impact

1. Introduction

Heavy metals are gaining attention as significant pollutants due to their toxicity problems in ecosystem at different levels [1]. Some of these metals are commonly known as pollutants that include arsenic, cadmium, copper, nickel and lead which pose threats of serious nature to aquatic environment and to the living organisms in the aquatic ecosystem [2]. Arsenic is a metalloid element that is abundant in the aquatic environment as a result of both natural and anthropogenic processes. It is a significant and ubiquitous environmental contaminant that causes health issues to all living organisms [3]. Arsenic in herbicides, fungicides, pesticides and rodenticides is the significant source of environmental contamination [4]. Arsenic mainly enters into the environment through two channels: (a) natural activities, and (b) man-made activities. Natural activities such as volcanic eruption, forest fires and weathering of rocks add a significant amount of arsenic in aquatic environment. While, man-made activities, such as different industries including paint, pharmaceutical, pesticide, detergent and electronic industries are the main source of arsenic discharge in water bodies [5]. Besides all these sources, the smelting and mining operations along with the domestic and agriculture run-off continuously add arsenic in natural waters [6]. In water, arsenic trivalent, arsenite is oxidized in water in the presence of dissolved oxygen and converted into arsenate that remain intact in sediments for long period of time and pose serious threats to aquatic fauna. Arsenic toxicity has been reported in many countries including China, Pakistan, Bangladesh, India and other South Asian countries along with many parts of the United States [7]. According to IARC, three chemical forms of arsenic are present: organic arsenic, inorganic arsenic and arsenic gas [8]. Fish is most sensitive bioindicator of pollution and cannot be safe from harmful impacts of these pollutants [9]. These have potential to induce biochemical and physiological changes which ultimately effect overall behavior, growth pattern and ultimately leads to death [10, 11]. Arsenic can enter in to fish body through oral cavity with contaminated food and absorption through skin and gills. Arsenic hasa tendency to accumulate in fish tissues and organs and cause serious damages to gills, gastrointestinal tract, kidneys, heart, brain and other organ. Such damages alter fish behavior, homoeostasis, hematology and biochemical mechanisms [12].

2. Sources of arsenic

In natural environment, arsenic is a common crystalline metalloid having characteristics of both metal and non-metal. It is the 14th and 20th abundant element in saltwater and earth crust respectively [13] Arsenic contamination occurs due to both natural (such as volcanic eruptions, rock weathering) and anthropogenic activities (such as the production of alloys, pesticides, glass, and medicinal items, *etc.*) [14].

2.1 Natural sources of arsenic

The presence of arsenic in natural water is influenced by the aquifer's local geology, hydrogeology and geochemical properties. Climate change and human activities also play a part and influence its presence. Natural sources of arsenic in water have been attributed to a variety of natural geochemical processes including oxidation of arsenic-bearing sulphides, desorption of arsenic from (hydro)oxides (e.g., iron, aluminum, and manganese oxides), reductive dissolution of arsenic-bearing iron (hydro)oxides, release of arsenic from geothermal water, and as well as leeching from sulphides.

2.1.1 Anthropogenic sources

Nonferrous metal mining and smelting, fossil fuel processing, combustion, wood preservation, pesticide production and its application in agricultural fields, municipal and industrial waste disposal and incineration are the main anthropogenic activities that may release arsenic into the environment [15, 16]. The majority of anthropogenic arsenic is released into the soil, primarily through pesticides or solid wastes. A significant amount, however, is also released into the air and water [17]. Arsenic, in its soluble forms, enters into the ground water and water bodies through runoff and leeching [18].

Mining tailing contain a significant amount of arsenic in the form of arsenopyrite, arsenian pyrite, arsenates and in association with iron oxyhydroxides. Arsenic can be produced by roasting arsenopyrite which is the most abundant ore mineral of arsenic, and by smelter dust of some metals such as gold, copper and lead [19]. Arsenic is found in approximately 11 million tonnes of copper and lead resources worldwide. In 2007, the total global production of arsenic trioxide was 59 thousand tonnes [20]. Arsenic is a highly toxic mineral found in the earth's crust that can enter the food chain through soil, water and plants. The main anthropogenic sources of arsenic in Canada are smelter and base-metal refinery facilities as well as thermal and power-generation stations. It was estimated that the Canadian base-metal smelters and refineries released approximately 15 tonnes of arsenic per year in liquid effluent, 310 tonnes in the atmosphere and 770 tonnes in solid waste [21]. The majority of this emissions (almost 90%) came from coal-fired power or thermal power-generation stations, as well as smelter and metallurgical facilities. Arsenic concentrations have been found to be elevated in the vicinity of these sources.

Another significant anthropogenic source of arsenic is the widespread use of arsenical wood preservatives. Chromated copper arsenate (CAA) is one of the most common wood preservatives used worldwide on large scale containing 34 percent arsenic content [22]. It has been estimated that a considerable amount of CAA (7800 to 78,000 mg/kg) remained in the treated wood [23]. Leaching of preservative components from in-service treated wood has thus been a source of arsenic in the environment. Furthermore, widespread arsenic contamination around wood preservation occurs as a result of raw material handling, spills, sludge deposition and dripping from freshly impregnated wood or rain water leaching from impregnated wood piles, particularly under low pH conditions, at these sites [24]. Exposure to sunlight and weathering both increase the rate of leaching from the treated wood. As a result, elevated arsenic levels have been found in soils surrounding treated woods [25, 26]. Arsenic is a widely used component in pesticides and most commonly available as lead arsenate, calcium aresenate, magnesium arsenate, zinc arsenate, arsenite and Paris green. These are commonly used in apple orchard of Canada [6].

2.1.2 Arsenic species in air

Methylated form of arsenic is mainly common in atmosphere. Arsenate is predominated form of arsenic and is most likely present near smelters, volcanic eruptions and coal burning sites [3]. Peat and landfills are major sources of volatile arsine in air. Some volatile species of arsenic are emitted into the air due to microbial activities in soil and water bodies which further oxidize and reactive with atmospheric sulfur oxide and ozone [6, 27]. This atmospheric arsenic reaches soil and water bodies by snow, dry deposition and through rain fall. Some of the atmospheric arsenic particles also combine with dust and may then be inhaled or ingested by living organisms [28]. In rural areas, the arsenic concentration in rain and snowfall are comparatively low (0.00003 mg/L), while its concentration around coal burning industries is relatively high (0.0005 mg/L) (**Figure 1**) [29].

3. Impacts of arsenic on fish

3.1 Prevalence of arsenic in the environment

Arsenic-exposed organisms may absorb arsenic by ingestion, inhalation and penetration through skin or mucous membranes which allows it to enter the cells

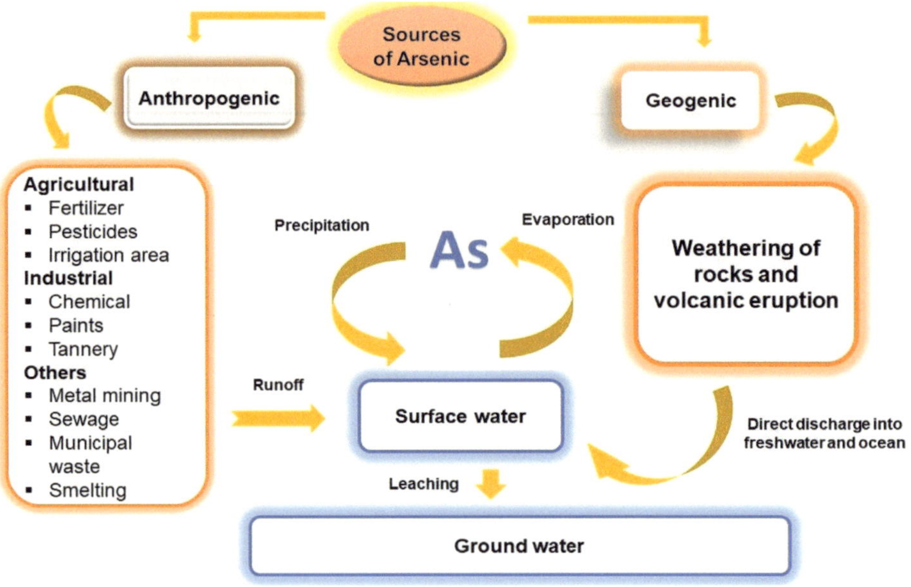

Figure 1.
Sources and routes of arsenic to approach water bodies. Arsenic has two main sources to enter water bodies: Anthropogenic and geogenic. Addition of arsenic in surface water bodies is by direct discharge. Gaseous arsenic from different sources also backs its way in to the water bodies through precipitation. Arsenic in soil and water enters the ground water through leeching.

mainly by active transport [30]. Inhaled arsenic induces severe biochemical and physiological changes such as poisoning, impaired growth and reproduction, immune system abnormalities, cell and tissue damage, oxidative stress and apoptosis in aquatic organisms.

3.2 Factors affecting toxicity of arsenic

3.2.1 Arsenic speciation

The toxic effect of Arsenic in the aquatic ecosystem depends on its form such as inorganic or organic and its level of oxidation [5]. Arsenite (As-III), arsine (As-III), Arsenate (AsV) and arsenic are the four oxidation states of arsenic [6]. Inorganic arsenic (iAs); As-III and As-V are prevalent forms in aquatic habitats. As-III is limited in extremely reduced conditions with a low redox potential while As-V is stable in oxygen-rich environments [31]. Inorganic metallic compounds are typically more hazardous than organometallic compounds. Likewise, inorganic Arsenic is more soluble in water than organic Arsenic it, thus, can accumulate in tissues more quickly [32]. Arsenic is converted into less- or non-toxic metabolites like arsenocholine (AsC) and arsenobetaine (AsB) for excretion when it is ingested by living organisms [33].

3.2.2 Biotic and abiotic factors

Many biotic and abiotic factors like exposure time, arsenic speciation, water temperature, pH, organic content, phosphate concentration, suspended particles, and presence of other chemicals and toxicants significantly alter the toxic and other effects of arsenic on aquatic life [34]. Median survival time of fishes usually reduces as the temperature and arsenic concentration increases. It is helpful to identify different arsenic species when investigating for arsenic exposure because they vary in their origin and toxicity [35].

By the biomethylation process, harmful inorganic arsenic is transformed into less toxic pentavalent (Met-A) forms such as monomethylarsonate (MMAv), dimethylarsenate (DMAv) and trimethylarsine (DMAIII). Nevertheless, monomethylarsenite (MMAIII), dimethylarsenite (DMAIII) and trimethylarsonic oxide (TMAOv) are more hazardous than inorganic arsenic and are produced through biomethylation. In general, AsB (arsenobetaine), arsenocholic (AsC), and DMAA make up around 85 to 90% of the arsenic found in edible parts of marine fish, whereas iAs (inorganic arsenic) species make up about 10%. Little is known about the types of arsenic found in freshwater fishes, but what is available indicates that AsB and DMAA are the dominant arsenic species in freshwater fishes [36].

3.2.3 Bioaccumulation

The process by which some toxic chemicals (heavy metals including arsenic and other toxicants) present in the environment accumulate in living organisms is known as bioaccumulation. Their absorption is considered passive, with diffusion gradients formed by metal adsorption or binding to tissue and cell surfaces [37]. Arsenic accumulation in tissue will depend on the rates at which various organs take in and eliminate arsenic [38]. Every organism metal concentration is determined by a variety of mechanisms, including its intake, excretion, storage, and transformation. Bioaccumulation differs among metal species and in fish species because of variations in their permeability, metabolic rates and the amount and types of metal binding ligands present at the organismic surface. Although accurate quantification of all these activities may not be necessary, understanding their proportional role in the overall pattern of metal turnover is sometimes the only way to evaluate tissue residue data.

The inhabitants of aquatic environments, such as fish, are unable to escape the negative impacts of arsenic [39]. The amount of toxic metal found in various fish organs is used as an indicator for the ecosystem metal contamination. This is thought as an important approach for highlighting the significance of higher metal levels in aquatic organisms [40]. Arsenic has a multidimensional impact on fish as they bioaccumulate in various tissues and can harm their immunological, respiratory, digestive, excretory, reproductive, neurological, and endocrine systems.

Mostly, arsenic accumulates in all of the fish critical organs. The most common site for the highest concentrations of arsenic in fish are the liver, kidney, and gills [41] as well as rarely the gut [42]. To determine the impact of arsenic contaminated water, several studies have been carried out. The gills and liver of tilapia have shown to accumulate significantly more arsenic thus the gill was noted to be most significantly impacted organ [43]. Great Slave Lake, Northwest Territories, Canada inhabitant fish arsenic concentration were determined in the gill, liver, muscle and skin of fishes *Clarias gariepinus* and *Labeo umbratus*. The liver was noted to be the most suspected organ [44]. In another study, the bioaccumulation of arsenic in the gill, liver, and intestine of freshwater fish *Oreochromis niloticus* was evaluated. The most affected organs among those mentioned above were the liver and gills. As the liver is the primary metabolic organ and the gills are always in constant contact with the environment, the gills were shown to be the most affected.

4. Effect of arsenic on behavioral changes

Fish exposed to chemicals exhibit quantifiable behavioral changes that provide unique information that cannot be discovered using traditional toxicological techniques [45]. The relationship between behavior and an organism physiology, ecology, and environment offers a special point of view [46]. Even a tiny quantity of some toxicants can make fish behave abnormally due to impaired sensory sensitivity. Numerous abnormal behaviors, including erratic movement, fast opercula movement, jumping out of the test medium, lateral swimming and loss of balance were observed by exposure to sodium arsenate [47]. Within a few minutes of exposure, especially at higher concentrations, 2.250 mgL^{-1} of sodium arsenate, the treated fish began to exhibit their first obvious responses. However, depending on the concentrations in the exposure medium, fish exposed to low concentrations, i.e., below 0.08 mgL^{-1} of sodium arsenate, had no or little behavioral alterations. Neurotoxic effects and sensory system irritability were the root causes of the aberrant behaviors. The fish avoidance responses to arsenic are indicated by their jumping and back-and-forth movements. The excessive mucus secretion was likely caused by sodium arsenate directly irritating the skin. The dysfunction of the nervous system may be the cause of lateral swimming and loss of balance [48].

5. Effect of arsenic on fish major organs and organ system

5.1 Effects on organs

Arsenic is hazardous to fish organs, including the skin, liver, kidney, lung, gastrointestinal tract and muscles [49]. Among them liver and kidneys, are essential organs in vertebrates that carry out detoxification processes, protein synthesis, homeostasis and excretion of nitrogenous waste respectively. Acute and sub-acute effects of arsenic may involve many organ systems particularly respiratory, cardiovascular, gastrointestinal, neurological and hematological systems. At 9.64 mgL^{-1} waterborne arsenic and 43.1–60 μgg^{-1} dietary arsenic concentration exposure anemia, liver degeneration and gallbladder inflammation were noted [50].

5.1.1 Skin

Fish skin serves as its outermost protective barrier. Because of the presence of club and mucous cells in the tissue, the skin and gill tissue of fish are keratinized and have a mucous covering [51]. It is susceptible to several water-dissolved toxins because of constant contact. Skin of *Clarias batrachus* (L.) exposed to 1 mgL^{-1} water born sodium arsenate induce significant damage, including enormous tear and wear, sloughing of epithelial cells and degeneration of club cells, whose contents seep out onto the body surface. This result in altered epidermal histomorphology. Mostly during exposure times, the mucous cells exhibit significant hyperplasia and hypertrophy [52]. Fish gills and skin develop a thick coating of slime as a defensive strategy. The protein, DNA and RNA contents of the epidermis also fluctuate periodically and independently [1]. With increased exposure time, damage to the epidermis becomes more prominent, and in certain areas, the majority of the cells in the bottom layer of the epidermis lost their integration and cell borders with adjacent cells [50]. Skin lipid and glycogen levels gradually decliness a result of exposure to arsenic, *Hetero pneustisfossilis* L. has been found to exhibit both hypo- and hyperpigmentation [53].

5.1.2 Gills

Gas exchange, ion control, and excretion of metabolic waste products are the three primary functions performed by gills. Consequently, by serving as a first barrier, gills can significantly contribute to the body defense against hazardous chemicals by reducing the number of poisonous compounds that are taken up by other organs [54]. Gills serve as the initial sites where waterborne contaminants are concentrated because of their constant exposure to the outside environment. One of the first signs of toxicant poisoning is respiratory discomfort. Fish are particularly susceptible to the toxicity of arsenic due to the high rate at which it is absorbed through the gills [55]. The presence of arsenic in gill tissue indicates that the gills were in direct touch with arsenic-contaminated water. The majority of research on arsenic toxicity in freshwater fish has focused on the effects of arsenic intake through the gills and nutritional absorption by fish that feed on benthic organisms [56]. Fish, upon arsenic exposure, show difficult breathing because coagulated mucus blocks the gills, and fish suffer direct damage from arsenic ions to their blood vessels which led to vascular collapse in the gills and anoxia [57]. After being exposed to sodium arsenite, *Oreochromis mossambicus* experienced the lifting of the lamellar epithelium, desquamation, edema, necrosis, aneurism, fusion of the secondary lamellae and hyperplasia of epithelial cells in the gills. Damage to gill tissues observed as a result of arsenic exposure can reduce oxygen consumption and also impaired osmoregulatory system. Hence, injury to the gills may reduce their ability to breathe effectively, altering the metabolic processes necessary for the growth and survival of fish [34].

5.1.3 Liver

By synthesizing proteins, detoxifying metabolites and aiding in digestion, the liver plays a significant role in metabolic regulation. Fish liver plays a crucial role in the absorption, bioaccumulation, biotransformation and elimination of arsenic [32]. Liver is the main target organ of arsenic poisoning. Arsenic is efficiently metabolized in fish tissue, particularly in the liver and gut and it tends to accumulate in fish such as the green sun fish and the *tilapia mossambica*. Its accumulation and detoxification lead to changes in the liver, including eosinophilic granules in the cytoplasm of hepatocytes, nuclear vacuolation, nuclear hypertrophy, and irregularly shaped nuclei [35]. The liver is very important for regulating metabolism through detoxification in the cytoplasm, brownish-yellow granules indicative of bile stagnation was seen. Melano-macrophages were recognized as compact cell clusters with granules that had a dark yellowish color. The severity of cellular rupture, bile stagnation and pyknotic nucleus increase with an increase in arsenic dosage [36]. The liver of the tilapia (*O. mossambicus*) exhibited congestion, focal lymphocytic and macrophage infiltration, cloudy swelling, shrinkage of hepatocytes, focal necrosis, vacuolization and dilation of sinusoids, vacuolar degeneration, and nuclear hypertrophy when exposed to various concentrations of arsenic. Chromosome fragmentation and the activation of certain proteins are brought on by sodium arsenite in the liver [37]. *Clarius batracus* exposed to arsenic had lower liver protein and glycogen levels and higher lipid contents [28]. In low-dose (50 μgL^{-1}) exposure to sodium arsenite, zebra fish liver proteome showed gender-specific responses [38]. With non-lethal exposure to arsenic, severe degenerative reactions in *Channa punctatus* hepatopancreas were seen [39]. In liver tissue of *C. batrachus* and *C. punctatus* hepatic pathologies were seen in the form of congestion, cloudy hepatocyte swelling, karyolysis, vacuolar degeneration and nuclear hypertrophy dilatation of sinusoids [40].

5.1.4 Kidney

Kidney along gills is primary pathways for waste excretion in fish body. When fish are exposed to arsenic contaminants, histopathological alterations occur. Upon arsenic exposure, kidney enzyme, glutathione decreases [40]. Upon non-lethal doses (3.8 mgL^{-1} and 7.6 mgL^{-1} arsenic) exposure *C. punctatus* (Bloch) suffer from shrinkage of glomerulus and increase in the Bowman's space consequently rise in urine quantity. Abnormalities in the renal tubule, such as necrotic and apoptotic cells, less intratubular space and more brush border cells with a higher height also occurred [41]. Arsenic accumulates in the head kidney of *C. batrachus* upon chronic exposure to arsenic, decreasing the number of head kidney macrophages and the head kidney somatic index while increasing hemosiderin accumulation. Head kidney macrophage showed significant endoplasmic reticulum, chromatin condensation, and a lack of nuclear membrane structural integrity. A considerable amount of superoxide anions and decreased generation of pro-inflammatory 'IL-1 β like' factors functioning as immunotoxic to fish and impairment in humoral response were detected [42]. In *C. batrachus* glomerulus shrinkage, vascular congestion, and raptured bowman's capsule were seen [43].

5.1.5 Gastrointestinal tract

The gastrointestinal system is the main pathway for dietary arsenic intake and absorption. Arsenic is delivered to the body organs via the circulatory system after being absorbed by the digestive tract [44]. Dietary exposure to arsenic has been demonstrated to damage the mucosal lining of the lake whitefish gastrointestinal system, causing mucosal sloughing and increased mucosal production [32]. Gastrointestinal disorders might result from acute arsenicexposure. Although the gastrointestinal effects are most noticeable immediately after ingesting arsenic, they can also develop with chronic exposure through other means. The primary gastrointestinal lesion appears to be increased small blood vessel permeability, which results in fluid loss and hypotension [45]. Exposure to a high concentration of arsenic, 20 mgL^{-1}, displayed disorganized, and consequent fusion of mucosa, lamina propria and edema, damaged serosa and degeneration [46]. Different Fish species exposed to various arsenic species included lake whitefish, walleye, northern pike, *Esoxlucius*, white sucker, *C. commersoni,* and longnose sucker (*Catostomus catostomus*). The concentrations of total arsenic and most arsenic species were maximum in the gastrointestinal tract in comparison to concentrations in liver and muscle [47]. The effects of arsenicals on tyrosine absorption by the winter flounder *Pseudopleuronectesamericanus*. Their findings showed that arsenicals inhibit Na-dependent tyrosine uptake [48].

5.1.6 Brain

Brain is extremely sensitive to arsenic because of its high rate of polyunsaturated fatty acids, oxygen consumption and extremely high rate of oxygen free radical formation without correspondingly large levels of arsenic [49]. It is thought that the arsenic poisoning altered the important biochemical components of the *Labeo rohita* brain tissues, such as proteins, lipids and nucleic acids, as indicated by the considerable variations in absorbance levels between the control and arsenic-intoxicated brain tissues [50]. The decrease in protein content upon arsenic exposure may be caused by decreased protein synthesis and by metal elements greater affinity for certain protein amino acid residues, which is regarded as the primary biochemical marker for early detection of stress. The conversion of protein into amino acid residues to expand the amino acid pool may possibly be the cause of the decreasing protein content. The modification of behavioral indicators and brain endonucleotidase activity is likewise caused by arsenic [51].

5.1.7 Muscles

The muscles, which make up to 80% of the fish itself, are what give the fish its swimming propulsion. The fish may move in any direction due to the muscle numerous orientations of arrangement (myomeres). Fish muscle, which makes up the majority of its bulk, is the part that people often eat [52]. The least amount of arsenic accumulated in the muscles across all experimental groups compared to other soft tissues. Muscle tissue does not directly come into touch with toxicants, so it is active detoxification site. As a result, arsenic is not transferred from other tissues to muscles. The least amount of arsenic has been found in the muscle of the *Mugil cephalus* [1]. Content of arsenic in muscle tissue increased significantly as fish grow. While continuous exposure to arsenic may change the size of muscle fibers, higher concentrations of metals were found in younger fish, which is typically due to the short residence times of these metals within the fish, associated with the increased metabolic rates in comparison to older organisms [53]. As exposure might start molecular alterations during embryogenesis that seem to cause abnormal muscle development [54]. *H. fossilis* fish exposed to 7 and 20 mgL^{-1} of arsenic had degeneration in their muscle bundles, specific regions of necrosis, atrophy, and vacuolar degeneration [46]. Fish muscle sampling revealed that arsenic speciation was not stable under various storage and sample preparation conditions.

5.1.8 Gonads

Fish reproduction was thought to be a reliable predictor of endocrine disruption caused by chemical substances, especially arsenic, in aquatic environments [55]. An earlier monitoring study in the Mekong Delta of Vietnam found a link between arsenic accumulation and gonad development in the catfish, *Pangasianodonhypophthalmus* [56]. Furthermore, after inducing spermatogenesis with human chorionic gonadotropin (hCG) through the synthesis of 11-KT, exposure of the testis to high concentrations of arsenic (100 μgL^{-1}) led to the production of reactive oxygen species (ROS) in the testis, which ultimately led to the apoptosis of germ cells. These results suggest that a low dose of arsenic decreases spermatogenesis by inhibition of steroidogenic enzyme activity and expression, whereas a large dose of same substance triggers oxidative stress-mediated germ cell death. In freshwater perch exposed to arsenite, ovarian degeneration was observed.

6. Biochemical and physiological changes

In aquatic medium, toxicants often exhibit their effects at the cellular or molecular level, which causes significant alterations in biochemical markers. Heavy metal pollution also has an impact on the body primary building blocks, such as lipids, proteins, and carbohydrates, which are crucial for building the body and generating energy [58]. Among these blood glucose level utilized as an indicator of environmental stress and showed how carbohydrate metabolism changed in the presence of hypoxia and stress. When Indian catfish, *C. batrachus*, were exposed to sub-lethal arsenic concentrations i.e., 0.50 μM, substantial cytological alterations and changes in serum biochemical markers were detected [59].

6.1 Carbohydrate

When fish are under stress, carbohydrates are main and immediate energy sources whereas, protein is spared. Changes in the levels of glucose, lactic acid and glycogen are among the effects of arsenic stress on fish carbohydrate metabolism. Among this blood glucose level was utilized as an indicator of environmental stress and showed how carbohydrate metabolism changed in the presence of hypoxia and stress [58]. Three important Indian carps, *Catla catla, L. rohita* and *Cirrhinus mrigala* were subjected to sub-lethal levels of toxic metals, including arsenic, and the results suggest that arsenic has a hyperglycemic impact [60]. Thus, in response to arsenic exposure significant glycogenolysis occur, which would raise blood sugar levels. As a result, a rise in blood glucose during sub-lethal exposure may be caused by decreased insulin secretion or gluconeogenesis to meet the increased metabolic demands imposed by arsenic stress. The large drop in plasma glucose levels after acute therapy may be due to arsenic-induced hypoxic conditions, which cause an excessive amount of stored carbs to be consumed.

6.1.1 Protein

Due to anoxic or hypoxic conditions, which increase carbohydrate consumption, heavy metal stress that affects glucose levels indicates a change in energy requirements and expenditure. As glycogen stores run short, tissue proteins use the process of deamination of amino acids to supply keto acids. Thus, a study of serum protein composition is needed to understand how energy requirements and expenditure change under metal stress. In order to determine fish overall nutritional status, estimation of total protein, albumin, and globulin in serum is of great diagnostic significance [61]. The ratio of albumin to globulin is a helpful measure for monitoring changes in the relative proportion of serum protein. Furthermore, hepatic tissue necrosis may cause a reduction in protein synthesis [62]. In *Cyprinus carpio* subjected to arsenic for 72 hours, serum albumin first decreased from 2 to 4 hours and then increased over the course of 72 hours whereas serum protein and globulin levels initially increased sharply from 2 to 20 hours and then decreased by 72 hours. When *C. gariepinus* was given arsenic, the serum total protein, albumin, and globulin levels rose [63].

6.1.2 Lipid

Lipid bilayers make up biomembranes, which also have different kinds of protein embedded in or attached to them. All biomembranes mostly consist of phospholipids. For the biomembrane to function properly, the lipid component composition must be maintained. Biomolecules including stored lipids, proteins, and carbs assist fish in dealing with stress. When a fish is under acute stress or toxicity, stored glycogen heals it; however, when a fish is exposed to arsenic continuously, the degree of the stress increases and lipids and proteins begin to play a role. Either the oxidation process or gradual saturation can mobilize lipids to supply the energy demand [64]. The integrity of the cell membrane is maintained by phospholipids and cholesterol. High-density lipoproteins (HDL), which are good cholesterol, and low-density lipoproteins (LDL), which are bad cholesterol, both exist in the body and aid in the removal of harmful cholesterol from the blood. The HDL cholesterol level should be as high as possible. As compared to LDL cholesterol, very-low-density lipoprotein (VLDL) is similar in that it is mostly made up of lipids with little protein. HDL is the primary serum lipoprotein in rainbow trout, followed by LDL and VLDL [65]. The primary form of reserve lipids, triacylglycerols, are mobilized prior to phospholipids during starvation. It is well-recognized that arsenic alter lipid levels. Hence, lipid profile analysis also functions as a biomarker for fish health. During arsenic intoxication serum total lipid levels in *C. gariepinus* significantly rose from 0.95 to 2.27 g/L [66]. Total cholesterol levels increased in *C. carpio* after 32 days of exposure to a combination of various metals [67]. As compared to fish from the reference population, *C. gariepinus* from Egypt's EL Rahawy drain, which is heavily contaminated with heavy metals (Cu, Fe, Pb, Cd, Mn, and arsenic), showed a considerable increase in blood total lipids, cholesterol, and triglyceride level [64].

7. Effect on oxidative status and other enzymes

Enzymes are biological macromolecules that regulate an organism metabolism. Much work has been conducted on how arsenic exposure affects the enzyme activity of certain fish [68]. Arsenic is absorbed through the gills, it has the potential to disturb the antioxidant system and impact the body reactions to oxidants by boosting glutamate cysteine ligase activity and glutathione levels Glucose-6-phosphate dehydrogenase (G6PDH) was significantly increased in fish gills upon exposure of arsenic, which modified antioxidant responses to an arsenic pro-oxidant challenge. As the produced nicotinamide adenine dinucleotide phosphate (NADPH) is an important element for the H_2O_2-scavenging pathway of cells [69] and for glutathione metabolism, there is indication that glucose-6-phosphate dehydrogenase, a major enzyme of the pentose phosphate pathway, has an important function in antioxidant systems [70]. Effect of sodium arsenate on *L. rohita* liver and muscle tissue and enzyme level was assessed and a significant decrease in acid phosphatase, glutamate-oxaloacetate transaminases, glutamate pyruvate transaminase and alkaline phosphatase enzyme activities were observed which revealed that arsenic causes disruptions in body metabolism [71]. Upon exposure of two non-lethal doses, an increase in the activity of antioxidant enzymes including superoxide dismutase (SOD), glutathione peroxidase ($GP_{x)}$ and catalase was noted in Indian cat fish, *C. batrachus*, however, after exposure, there was a significant drop in glutathione reductase (GR) activity, demonstrating the oxidative stress in fish. As a result of the excessive H_2O_2 generation, arsenic toxicity in *C. batrachus* may be targeted at the peroxisomal metabolizing enzymes. Reactive oxygen species production is linked to arsenic toxicity, which can seriously harm or damage the neurological system [72]. Enzyme activities are known as sensitive biochemical markers and are frequently used in aquatic toxicology to evaluate the health of the organism [73]. Alanine aminotransferase (ALT) and aspartate aminotransferase (AST) and are two of the many enzymes that are evaluated, and they are frequently used to identify tissue damage brought on by toxicants [74]. *C. batrachus* exposed to arsenic at sub-lethal concentrations higher level of aspartate aminotransferase and alanine aminotransferase was noted [59]. Toxicity is linked to the creation of reactive oxygen species, due to its capacity to bind to SH groups, arsenic can inhibit the activity of numerous enzymes, including those involved in cellular glucose absorption, gluconeogenesis and glutathione synthesis, fatty acid oxidation (77). Many blood serum enzymes have been used as markers for hepatic dysfunction and injury. ALT and AST are enzymes that employed, and they are frequently used to identify tissue damage induced by toxicants [74]. In *L. rohita* exposed to arsenic, found a considerable reduction in liver ALT and AST which implies a major decline in the structure and function of cell organelles such the endoplasmic reticulum and the mitochondria [71]. Also, it was hypothesized that the cells utilize the phosphate-like substance, arsenate, for energy and signaling. Arsenic has the ability to inhibit the synthesis of energy and normal cell signaling by replacing phosphate in enzymes or signaling proteins (77). Delta aminolevulinic acid dehydratase (ALAD) levels were assessed to record the disruption in hemoglobin production and erythrocyte alterations. Arsenic exposure decreased the activity of ALAD in the blood by 62%. 2,3-Dimercaptosuccinic acid andALAD levels could be returned to control levels by using DMSA alone or in combination with N-acetylcysteine just the mixture of the two was able to raise RBC glutathione levels in those who were not exposed to arsenic [75].

8. Hematological changes

The hematopoietic system is influenced by short- and long-term arsenic exposure. Hematological profiles of fish are commonly used to detect the environmental contamination in aquatic ecosystem [76]. Different fish blood parameters are used to determine the effects of sub-lethal arsenic. Hematological and biochemical examinations of blood parameters in fish exposed to pollutants are crucial for determining the animal structural and functional status [77]. Red and white blood cell formation may be reduced as a result of arsenic exposure [78]. Leukocyte numbers were reduced as a result of chronic arsenic exposure which had an impact on the structure of the head kidney [50]. Arsenic induces changes to hematological markers and oxidative stress in the fish liver [79]. Numerous studies [80–82] have noted an anemic state of the fish during acute and sub-lethal treatment, which led to a low level of hemoglobin (Hb) in the arsenate-treated fish. Another potential explanation is that the toxicity of arsenic may inhibit erythropoiesis due to its effect on membranes. It was also that the fish exposed to toxicants had a lower amount of red blood cells [83]. In *Oryctolaguscuniculus*, that had been exposed to waterborne arsenic, a decrease in hemoglobin and packed cell volume was noted [84]. WBC count in *O. mykiss* exposed to arsenic was found to have decreased significantly due to reduction in lymphocytes [84]. It is commonly stated that changes in leukocyte counts upon exposure to contaminants may be linked to a decline in the fish nonspecific immunity. The control of immunological function by leukocytes occurs in a wide range of species, and the development of more WBCs in animals under stress suggests a defense mechanism against stress [85]. Chronic arsenic exposure changed the shape of the head kidney and was linked to a decrease in leukocytes [43]. The formation of reticular tissues, edematous growth, and lower leukocyte counts in the kidney and spleen, together with the establishment of an unusual lymphocyte population, indicate that arsenic probably impacts the process of lymphopoiesis and blast development in *C. batrachus*. Another cause could be because arsenic effect prevents WBCs from maturing and releasing themselves from tissue reservoirs. Moreover, exposure of arsenic in *C. carpio* led to reduced numbers of granulocyte, hemoglobin, erythrocyte and hematocrit when compared to the control group [86].

9. Immunotoxic effects

Arsenic as an immunotoxic substance has an impact on a variety of immunological responses, including altering co-receptor expression, lowering delayed hypersensitivity reactions, reducing mitogen-activated T-cell proliferation, production of freer intracellular $Ca^{2+}$and releasing lymphokines [87]. The primary immune-competent fish organs are the head kidney, spleen and thymus [88]. The head kidney macrophages (HKM) are essential for the activation of fish innate immunity, and arsenic-induced macrophage mortality is going to impair the immunological system of the exposed fish. Arsenic accumulates in fish liver and kidney and it can impair the fish immune system by decreasing the synthesis of antibodies and cytokines [89]. Sub-lethal fish become immunocompromised and vulnerable to infections as a result of arsenic exposure, which alters the functional arms of their innate and acquired immune systems [90]. In fish, adaptive immunity develops later, therefore the impact of ecotoxins on innate immunity may be more substantial [91]. In zebra fish, the system expressed crucial antiviral genes and generated enough tumor necrosis factor (TNF-α) to fall within the range of arsenic [92]. Arsenic has a significant impact on the immune system in fish, with the two immunologically significant organs, the head kidney and spleen, responding to its toxic effects in various ways. Arsenic caused a drop in both T- and B-lymphocytes cell responses in the head kidney and spleen, although its effects seem to be more prominent on the B-cells. The phagocytic capability of macrophages was similarly impacted by fish exposure to different arsenic concentrations, which helped in the spread and duration of bacterial and viral infections [42].

10. Cytogenotoxic effects

Arsenic is recognized as a possible sulfhydryl-reactive substance that may bind to and aggregate many cell surface proteins [93]. Arsenic increases nitric oxide generation at the level of transcriptional activation along with inducing ribosylation, polyadenosine diphosphate, DNA strand breaks, depletion of nicotinamide adenine dinucleotide and the development of micronuclei, like other oxygen radical-producing stressors [94]. Cell death may be caused by the accumulation of cellular proteins, the generation of reactive oxygen species, or the stimulation of protein tyrosine kinases by arsenic [36]. Furthermore, denaturing of biological enzymes and changing gene regulation are toxic consequences of inorganic arsenic. For the purpose of researching the cytotoxicity of various arsenic compounds, fish cell lines may be used as sensitive substitutes for entire fish. In JF cells, arsenite may cause apoptosis by the induction of oxidative stress, however in TO-2 fish cell lines, it disrupts the cell cycle without the induction. Arsenic may impair cell division by disrupting the spindle apparatus [95]. In addition, it causes sister chromatid exchange, the formation of micronuclei, DNA-protein crosslinking, and different types of mutations [96]. In reaction to arsenic, a duration- and dose-dependent increase in the formation of micronuclei in the gill cells of *Danio rerio* was reported [97]. It is known to prevent DNA repair [98] and even worsen the effects of other mutagenic substances [59], making fish more susceptible to a variety of disorders. There are abundant evidences thatarsenic can interfere with gene expression, especially due to its impact on signal transduction [59]. The DNA-binding abilities of the transcription factors NFkβ, and AP-1, i.e., 'activator protein 1', were demonstrated to be affected by low

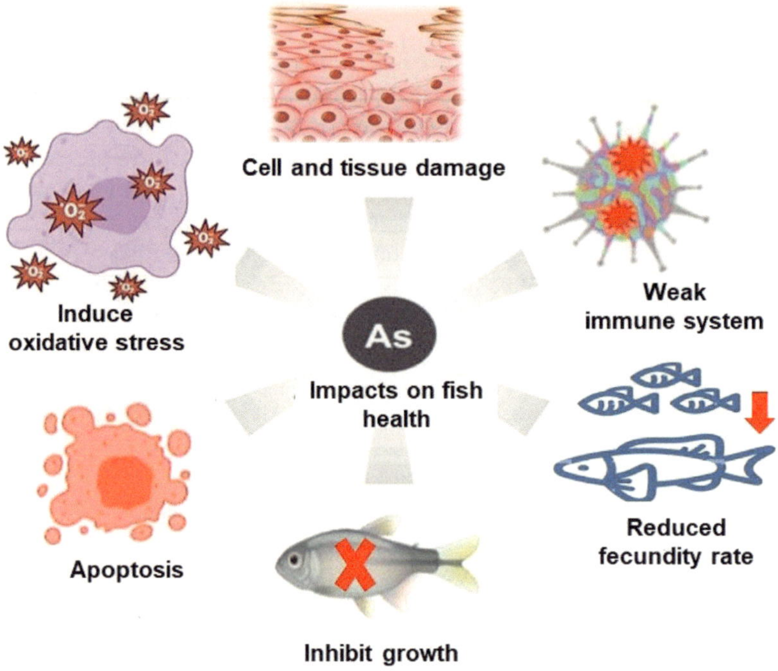

Figure 2.
Impact of arsenic on different organs of fish. Induction of oxidative stress, inhibition of fish growth, weakness of the immune system, reduced fecundity rate, damaging of the cell and tissues and enhancement of the cell death through apoptosis are the effects of arsenic toxicity in fish.

arsenicconcentrations, which increased gene expression and induced cell proliferation [99]. High concentrations of arsenic, however, may reduce NFjβ activation, restrict cell growth and cause apoptosis [100]. The co-mutagenicity and possibly the co-carcinogenicity of arsenic may be primarily mediated by the repair inhibition. Arsenic exposure may lead to DNA hypomethylation as a result of ongoing methyl depletion, which supports abnormal gene expression (**Figure 2**).

Author details

Ayesha Malik[1], Fakhira Khalid[2], Nigah Hidait[2], Khalid Mehmood Anjum[2], Saima Mahad[3], Abdul Razaq[4], Hamda Azmat[1*] and Muhammad Bilal Bin Majeed[5*]

1 Department of Fisheries and Aquaculture, University of Veterinary and Animal Sciences, Lahore, Pakistan

2 Department of Wildlife and Ecology (Zoology), University of Veterinary and Animal Sciences, Lahore, Pakistan

3 Department of Animal Nutrition, University of Veterinary and Animal Sciences, Lahore, Pakistan

4 Department of Biological Sciences, University of Veterinary and Animal Sciences, Lahore, Pakistan

5 Department of Animal Breeding and Genetics (Section Genetics), University of Veterinary and Animal Sciences, Lahore, Pakistan

*Address all correspondence to: hamda.azmat@uvas.edu.pk
and bilal.majeed@uvas.edu.pk

References

[1] Jaishankar M, Tseten T, Anbalagan N, Mathew BB, Beeregowda KN. Toxicity, mechanism and health effects of some heavy metals. Interdisciplinary Toxicology. 2014;**7**(2):60

[2] Lambert TW, Lane S. Lead, arsenic, and polycyclic aromatic hydrocarbons in soil and house dust in the communities surrounding the Sydney, Nova Scotia, tar ponds. Environmental Health Perspectives. 2004;**112**(1):35-41

[3] Rossman TG. Mechanism of arsenic carcinogenesis: An integrated approach. Mutation Research/Fundamental and Molecular Mechanisms of Mutagenesis. 2003;**533**(1-2):37-65

[4] Nickson R, McArthur J, Burgess W, Ahmed KM, Ravenscroft P, Rahmanñ M. Arsenic poisoning of Bangladesh groundwater. Nature. 1998;**395**(6700):338

[5] Canivet V, Chambon P, Gibert J. Toxicity and bioaccumulation of arsenic and chromium in epigean and hypogean freshwater macroinvertebrate. Archives of Environmental Contamination and Toxicology. 2001;**40**:345-354

[6] Brammer H, Ravenscroft P. Arsenic in groundwater: A threat to sustainable agriculture in South and South-East Asia. Environment International. 2009;**35**(3):647-654

[7] Yoshida T, Yamauchi H, Sun GF. Chronic health effects in people exposed to arsenic via the drinking water: Dose–response relationships in review. Toxicology and Applied Pharmacology. 2004;**198**(3):243-252

[8] IARC Working Group on the Evaluation of Carcinogenic Risks to Humans. Arsenic, Metals, Fibres and Dusts. Lyon (FR): International Agency for Research on Cancer. (IARC Monographs on the Evaluation of Carcinogenic Risks to Humans, No. 100C.) ARSENIC AND ARSENIC COMPOUNDS. 2012. Available from: https://www.ncbi.nlm.nih.gov/books/NBK304380/

[9] Suganya A, Murugan K, Kovendan K, Mahesh Kumar P, Hwang JS. Green synthesis of silver nanoparticles using Murraya koenigii leaf extract against Anopheles stephensi and Aedes aegypt. Parasitology Research. 2013;**112**:1385-1397

[10] Beyers DW, Rice JA, Clements WH, Henry CJ. Estimating physiological cost of chemical exposure: Integrating energetics and stress to quantify toxic effects in fish. Canadian Journal of Fisheries and Aquatic Sciences. 1999;**56**(5):814-822

[11] Guba M, von Breitenbuch P, Steinbauer M, Koehl G, Flegel S, Hornung M, et al. Rapamycin inhibits primary and metastatic tumor growth by antiangiogenesis: Involvement of vascular endothelial growth factor. Nature Medicine. 2002;**8**(2):128-135

[12] Lavanya D, Rani DK. Analysis of feature selection with classification: Breast cancer datasets. Indian Journal of Computer Science and Engineering (IJCSE). 2011;**2**(5):756-763

[13] Popovic A, Djordjevic D, Polic P. Trace and major element pollution originating from coal ash suspension and transport processes. Environment International. 2001;**26**(4):251-255

[14] Bhattacharya P, Mukherjee AB, Jacks G, Nordqvist S. Metal contamination at a wood preservation site:

Characterisation and experimental studies on remediation. Science of the Total Environment. 2002;**290**(1-3):165-180

[15] Nasser NA, Patterson RT, Roe HM, Galloway JM, Falck H, Sanei H. Use of Arcellinida (testate lobose amoebae) arsenic tolerance limits as a novel tool for biomonitoring arsenic contamination in lakes. Ecological Indicators. 2020;**113**:106177

[16] Slimak M, Delos C. Environmental pathways of exposure to 129 priority pollutants. Journal of Toxicology: Clinical Toxicology. 1983;**21**(1-2):39-63

[17] Duker AA, Carranza E, Hale M. Arsenic geochemistry and health. Environment International. 2005;**31**(5):631-641

[18] Pongratz R. Arsenic speciation in environmental samples of contaminated soil. Science of the Total Environment. 1998;**224**:133-141

[19] Smedley PL, Kinniburgh DG. A review of the source, behaviour and distribution of arsenic in natural waters. Applied Geochemistry. 2002;**17**(5):517-568

[20] Fields S. Caution–children at play: How dangerous is CCA? Environmental Health Perspectives. 2001;**109**(6):A262-A269

[21] Lund U, Fobian A. Pollution of two soils by arsenic, chromium and copper, Denmark. Geoderma. 1991;**49**(1-2):83-103

[22] Stephens RW, Brudermann GE, Morris PI, Hollick MS, Chalmers JD. Value assessment of the Canadian pressure treated wood industry Report to Canadian Forest Service; 1994

[23] Zagury GJ, Samson R, Deschênes L. Occurrence of metals in soil and ground water near chromated copper arsenate–treated utility poles. Journal of Environmental Quality. 2003;**32**(2):507-514

[24] Frank R, Braun HE, Ishida K, Suda P. Persistent organic and inorganic pesticide residues in orchard soils and vineyards of Southern Ontario. Canadian Journal of Soil Science. 1976;**56**(4):463-484

[25] Cullen WR, Reimer KJ. Arsenic speciation in the environment. Chemical Reviews. 1989;**89**:713-764

[26] Challenger F. Biological methylation. Chemical Reviews. 1945;**36**(3):315-361

[27] Rasmussen PE, Subramanian KS, Jessiman BJ. A multi-element profile of house dust in relation to exterior dust and soils in the city of Ottawa, Canada. Science of the Total Environment. 2001;**267**(1-3):125-140

[28] Andreae MO. Arsenic in rain and the atmospheric mass balance of arsenic. Journal of Geophysical Research Oceans. 1980;**85**(C8):4512-4518

[29] Genchi G, Lauria G, Catalano A, Carocci A, Sinicropi MS. Arsenic: A review on a great health issue worldwide. Applied Sciences. 2022;**12**(12):6184. DOI: 10.3390/app12126184

[30] Masuda H. Arsenic cycling in the Earth's crust and hydrosphere: Interaction between naturally occurring arsenic and human activities. Progress in Earth and Planetary Science. 2018;**5**(1):1-1. DOI: 10.1186/s40645-018-0224-3

[31] Mandal BK, Suzuki KT. Arsenic round the world: A review. Talanta. 2002;**58**(1):201-235

[32] Camacho J, de Conti A, Pogribny IP, Sprando RL, Hunt PR. Assessment

of the effects of organic vs. inorganic arsenic and mercury in Caenorhabditis elegans. Current Research in Toxicology. 2022;**1**(3):100071

[33] Byeon E, Kang HM, Yoon C, Lee JS. Toxicity mechanisms of arsenic compounds in aquatic organisms. Aquatic Toxicology. 2021;**237**:105901

[34] Ardini F, Dan G, Grotti M. Arsenic speciation analysis of environmental samples. Journal of Analytical Atomic Spectrometry. 2020;**35**(2):215-237

[35] Chen SJ, Yan XJ, Chen Z. Arsenic in Nature. In: Kretsinger RH, Uversky VN, Permyakov EA, editors. Encyclopedia of Metalloproteins. New York, NY: Springer; 2013. DOI: 10.1007/978-1-4614-1533-6_489

[36] Neff JM. Ecotoxicology of arsenic in the marine environment. Environmental Toxicology and Chemistry: An International Journal. 1997;**16**(5):917-927

[37] Kumaresan M, Riyazuddin P. Overview of speciation chemistry of arsenic. Current Science. 2001;**10**:837-846

[38] Barral-Fraga L, Barral MT, MacNeill KL, Martiñá-Prieto D, Morin S, Rodríguez-Castro MC, et al. Biotic and abiotic factors influencing arsenic biogeochemistry and toxicity in fluvial ecosystems: A review. International Journal of Environmental Research and Public Health. 2020;**17**(7):2331. DOI: 10.3390/ijerph17072331

[39] Orloff K, Mistry K, Metcalf S. Biomonitoring for environmental exposures to arsenic. Journal of Toxicology and Environmental Health, Part B. 2009;**12**(7):509-524

[40] Šlejkovec Z, Bajc Z, Doganoc DZ. Arsenic speciation patterns in freshwater fish. Talanta. 2004;**62**(5):931-936

[41] Bryan GW, Darracott A. Bioaccumulation of marine pollutants. Philosophical Transactions of the Royal Society of London. B, Biological Sciences. 1979;**286**(1015):483-505

[42] Amiard-Triquet C, Amiard JC. Influence of ecological factors on accumulation of metal mixtures. In: Langston WJ, Bebianno MJ, editors. Metal Metabolism in Aquatic Environments. Boston, MA: Springer; 1998. DOI: 10.1007/978-1-4757-2761-6_11

[43] Mendil D, Uluözlü ÖD, Hasdemir E, Tüzen M, Sari H, Suicmez M. Determination of trace metal levels in seven fish species in lakes in Tokat, Turkey. Food Chemistry. 2005;**90**(1-2):175-179

[44] Squadrone S, Prearo M, Brizio P, Gavinelli S, Pellegrino M, Scanzio T, et al. Heavy metals distribution in muscle, liver, kidney and gill of European catfish (Silurus glanis) from Italian Rivers. Chemosphere. 2013;**90**(2):358-365

[45] Farag AM, Stansbury MA, Bergman HL, Hogstrand C, MacConnell E. The physiological impairment of free-ranging brown trout exposed to metals in the Clark Fork River, Montana. Canadian Journal of Fisheries and Aquatic Sciences. 1995;**52**(9):2038-2050

[46] Wepener V, Van Vuren JH, Du Preez HH. Uptake and distribution of a copper, iron and zinc mixture in gill, liver and plasma of a freshwater teleost, Tilapia sparrmanii. Water SA. 2001;**27**(1):99-108

[47] Chetelat J, Cott PA, Rosabal M, Houben A, McClelland C, Belle Rose E, et al. Arsenic bioaccumulation in subarctic fishes of a mine-impacted bay on Great Slave Lake, Northwest Territories, Canada. PLoS One. 2019;**14**(8):e0221361

[48] Saglio P, Trijasse S. Behavioral responses to atrazine and diuron in goldfish. Archives of Environmental Contamination and Toxicology. 1998;**35**:484-491

[49] Kumari B, Kumar V, Sinha AK, Ahsan J, Ghosh AK, Wang H, DeBoeck G. Toxicology of arsenic in fish and aquatic systems. Environmental chemistry letters. 2017;**15**:43-64

[50] Baldissarelli LA, Capiotti KM, Bogo MR, Ghisleni G, Bonan CD. Arsenic alters behavioral parameters and brain ectonucleotidases activities in zebrafish (Danio rerio). Comparative Biochemistry and Physiology Part C: Toxicology & Pharmacology. 2012;**155**(4):566-572

[51] Tolins M, Ruchirawat M, Landrigan P. The developmental neurotoxicity of arsenic: Cognitive and behavioral consequences of early life exposure. Annals of Global Health. 2014;**80**(4):303-314

[52] Khan MI, Ahmad MF, Ahmad I, Ashfaq F, Wahab S, Alsayegh AA, et al. Arsenic exposure through dietary intake and associated health hazards in the Middle East. Nutrients. 2022;**14**(10):2136. DOI: 10.3390/nu14102136

[53] Singh AK, Banerjee TK. Toxic effects of sodium arsenate (Na2HAsO4x7H2O) on the skin epidermis of air-breathing catfish *Clarias batrachus* (L.). Veterinarski Arhiv. 2008;**78**(1):73-88

[54] Cockell KA, Hilton JW, Bettger WJ. Hepatobiliary and hematological effects of dietary disodium arsenate heptahydrate in juvenile rainbow trout (*Oncorhynchus mykiss*). Comparative Biochemistry and Physiology Part C: Comparative Pharmacology. 1992;**103**(3):453-458

[55] Rakers S, Gebert M, Uppalapati S, Meyer W, Maderson P, Sell AF, et al. 'Fish matters': The relevance of fish skin biology to investigative dermatology. Experimental Dermatology. 2010;**19**(4):313-324

[56] Kumar R, Banerjee TK. Analysis of arsenic bioaccumulation in different organs of the nutritionally important catfish, *Clarias batrachus* (L.) exposed to the trivalent arsenic salt, sodium arsenite. Bulletin of Environmental Contamination and Toxicology. 2012:445-449

[57] Kumar R, Banerjee TK. Analysis of arsenic bioaccumulation in different organs of the nutritionally important catfish, Clarias batrachus (L.) exposed to the trivalent arsenic salt, sodium arsenite. Bulletin of environmental contamination and toxicology. 2012;**89**:445-9

[58] Mdegela R, Myburgh J, Correia D, Braathen M, Ejobi F, Botha C, et al. Evaluation of the gill filament-based EROD assay in African sharptooth catfish (*Clarias gariepinus*) as a monitoring tool for waterborne PAH-type contaminants. Ecotoxicology. 2006;**15**:51-59

[59] Lima AA, Furtado C, Valduriez P, Mattoso M. Parallel OLAP query processing in database clusters with data replication. Distributed and Parallel Databases. 2009;**25**:97-123

[60] Pedlar RM, Ptashynski MD, Wautier KG, Evans RE, Baron CL, Klaverkamp JF. The accumulation, distribution, and toxicological effects of dietary arsenic exposure in lake whitefish (Coregonus clupeaformis) and lake trout (Salvelinus namaycush). Comparative Biochemistry and Physiology Part C Toxicology & Pharmacology. 2002;**131**(1):73-91

[61] Mondal K, Samanta S. A review on arsenic contamination in fresh water fishes of West Bengal. Journal of Global Biosciences. 2015;**4**(5):2369-2374

[62] Ahmed MK, Habibullah-Al-Mamun M, Parvin E, Akter MS, Khan MS. Arsenic induced toxicity and histopathological changes in gill and liver tissue of freshwater fish, tilapia (*Oreochromis mossambicus*). Experimental and Toxicologic Pathology. 2013;**65**(6):903-909

[63] Caumette G, Koch I, Estrada E, Reimer KJ. Arsenic speciation in plankton organisms from contaminated lakes: Transformations at the base of the freshwater food chain. Environmental Science & Technology. 2011;**45**(23):9917-9923

[64] Hossain K, Akhand AA, Kato M, Du J, Takeda K, Wu J, et al. Arsenite induces apoptosis of murine T lymphocytes through membrane raft-linked signaling for activation of c-Jun amino-terminal kinase. The Journal of Immunology. 2000;**165**(8):4290-4297

[65] Das S, Unni B, Bhattacharjee M, Wann SB, Rao PG. Toxicological effects of arsenic exposure in a freshwater teleost fish, *Channa Punctatus*. African Journal of Biotechnology. 2012;**11**(19):4447-4454

[66] Carlson P, Smalley DM, Van Beneden RJ. Proteomic analysis of arsenic-exposed zebrafish (Danio rerio) identifies altered expression in proteins involved in fibrosis and lipid uptake in a gender-specific manner. Toxicological Sciences. 2013;**134**(1):83-91

[67] Das T, Goswami M. Histopathological and ultrastructural changes in the gill and liver of fresh water fish *Channa punctatus* exposed to sodium arsenite. Bioscience Biotechnology Research Communications. 2018;**11**(3):434-441

[68] Allen T, Singhal R, Rana SV. Resistance to oxidative stress in a freshwater fish *Channa punctatus* after exposure to inorganic arsenic. Biological Trace Element Research. 2004;**98**:63-72

[69] Roy S, Bhattacharya S. Arsenic-induced histopathology and synthesis of stress proteins in liver and kidney of *Channa punctatus*. Ecotoxicology and Environmental Safety. 2006;**65**(2): 218-229

[70] Ghosh D, Datta S, Bhattacharya S, Mazumder S. Long-term exposure to arsenic affects head kidney and impairs humoral immune responses of *Clarias batrachus*. Aquatic Toxicology. 2007;**81**(1):79-89

[71] Datta S, Ghosh D, Saha DR, Bhattacharaya S, Mazumder S. Chronic exposure to low concentration of arsenic is immunotoxic to fish: Role of head kidney macrophages as biomarkers of arsenic toxicity to *Clarias batrachus*. Aquatic Toxicology. 2009;**92**(2):86-94

[72] Gelberg H. Pathophysiological mechanisms of gastrointestinal toxicity. Comprehensive Toxicology. 2018:139-178. DOI: 10.1016/B978-0-12-801238-3.10923-7

[73] Haque M, Roy SK. Acute effects of arsenic on the regulation of metabolic activities in liver of fresh water fishes (Taki) during cold acclimation. Jordan Journal of Biological Sciences. 2012;**5**(2):91-97

[74] Begum A, Mustafa AI, Amin MN, Banu N, Chowdhury TR. Accumulation and histopathological effects of arsenic in tissues of shingi fish (stinging catfish) Heteropneustes fossilis (Bloch, 1794). Journal of the Asiatic Society of Bangladesh, Science. 2013;**39**(2):221-230

[75] de Rosemond S, Xie Q, Liber K. Arsenic concentration and speciation in five freshwater fish species from Back Bay near Yellowknife, NT, Canada. Environmental Monitoring and Assessment. 2008;**147**:199-210

[76] Chauncey B, Schmid EC, Goldstein L. Arsenical and mercurial inhibition of tyrosine transport by the flounder intestine. Journal of Toxicology and Environmental Health, Part A Current Issues. 1988;**23**(2):257-265

[77] Ng SC, Furman R, Axelsen PH, Shchepinov MS. Free radical chain reactions and polyunsaturated fatty acids in brain lipids. ACS Omega. 2022;7(29):25337-25345. DOI: 10.1021/acsomega.2c02285

[78] Palaniappan PR, Vijayasundaram V. The effect of arsenic exposure and the efficacy of DMSA on the proteins and lipids of the gill tissues of *Labeo rohita*. Food and Chemical Toxicology. 2009;**47**(8):1752-1759

[79] Palaniappan PL, Vijayasundaram V. The bioaccumulation of arsenic and the efficacy of Meso-2, 3-dimercaptosuccinic acid in the selected organ tissues of Labeo rohita fingerlings using inductively coupled plasma-optical emission spectrometry. WASJ. 2009;**6**:1247-1254

[80] Maher W, Goessler W, Kirby J, Raber G. Arsenic concentrations and speciation in the tissues and blood of sea mullet (Mugil cephalus) from Lake Macquarie NSW, Australia. Marine Chemistry. 1999;**68**(1-2):169-182

[81] D'Amico AR, Gibson AW, Bain LJ. Embryonic arsenic exposure reduces the number of muscle fibers in killifish (Fundulus heteroclitus). Aquatic Toxicology. 2014;**146**:196-204

[82] Dubińska-Magiera M, Daczewska M, Lewicka A, Migocka-Patrzałek M, Niedbalska-Tarnowska J, Jagla K. Zebrafish: A model for the study of toxicants affecting muscle development and function. International Journal of Molecular Sciences. 2016;**17**(11):1941. DOI: 10.3390/ijms1711194

[83] Arcand-Hoy LD, Benson WH. Fish reproduction: An ecologically relevant indicator of endocrine disruption. Environmental Toxicology and Chemistry An International Journal. 1998;**17**(1):49-57

[84] Celino FT, Yamaguchi S, Miura C, Miura T. Arsenic inhibits in vitro spermatogenesis and induces germ cell apoptosis in Japanese eel (Anguilla japonica). Reproduction. 2009;**138**(2):279

[85] Javed M, Usmani N. Stress response of biomolecules (carbohydrate, protein and lipid profiles) in fish *Channa punctatus* inhabiting river polluted by thermal power plant effluent. Saudi Journal of Biological Sciences. 2015;**22**(2):237-242

[86] Datta S, Saha DR, Ghosh D, Majumdar T, Bhattacharya S, Mazumder S. Sub-lethal concentration of arsenic interferes with the proliferation of hepatocytes and induces in vivo apoptosis in Clarias batrachus L. Comparative Biochemistry and Physiology Part C, Toxicology & Pharmacology. 2007;**145**(3):339-349

[87] Garg S, Gupta RK, Jain KL. Sublethal effects of heavy metals on biochemical composition and their recovery in Indian major carps. Journal of Hazardous Materials. 2009;**163**(2-3):1369-1384

[88] Abdel-Tawwab M, Abdel-Rahman AM, Ismael NE. Evaluation of commercial live bakers' yeast, Saccharomyces cerevisiae as a growth and immunity promoter for fry Nile tilapia, *Oreochromis niloticus* (L.) challenged

in situ with Aeromonas hydrophila. Aquaculture. 2008;**280**(1-4):185-189

[89] Jacobs MN, Covaci A, Schepens P. Investigation of selected persistent organic pollutants in farmed Atlantic salmon (Salmo salar), salmon aquaculture feed, and fish oil components of the feed. Environmental Science & Technology. 2002;**36**(13):2797-2805

[90] Ghaffar A, Hussain R, Aslam M, Abbas G, Khan A. Arsenic and urea in combination alters the hematology, biochemistry and protoplasm in exposed rahu fish (*Labeo rohita*)(Hamilton, 1822). Turkish Journal of Fisheries and Aquatic Sciences. 2016;**16**(2):289-296. DOI: 10.4194/1303-2712-v16_2_09

[91] Abhay DS. Comparative study of cholesterol alterations in a freshwater teleost fish, Amblypharyngodon mola exposure to heavy metals. The Bioscan Journal of Life Sciences. 2013;**8**(3):1001-1003

[92] Perrier H, Perrier C, Peres G, Gras J. The lipoproteins of the plasma of the rainbow trout (Salmo gairdnerii Richardson): Immunoelectrophoresis, selective precipitation and lipid composition. Comparative Biochemistry and Physiology Part B, Comparative Biochemistry. 1979;**62**(3):245-248

[93] Abdel-Tawwab M, Mousa MA, Abbass FE. Growth performance and physiological response of African catfish, *Clarias gariepinus* (B.) fed organic selenium prior to the exposure to environmental copper toxicity. Aquaculture. 2007;**272**(1-4):335-345

[94] Vinodhini R, Narayanan M. The impact of toxic heavy metals on the hematological parameters in common carp (*Cyprinus carpio* L.). Journal of Environmental Health Science & Engineering. 2009;**6**(1):23-28

[95] Morales AE, Pérez-Jiménez A, Hidalgo MC, Abellán E, Cardenete G. Oxidative stress and antioxidant defenses after prolonged starvation in Dentex dentex liver. Comparative Biochemistry and Physiology Part C, Toxicology & Pharmacology. 2004;**139**(1-3):153-161

[96] Bagnyukova TV, Luzhna LI, Pogribny IP, Lushchak VI. Oxidative stress and antioxidant defenses in goldfish liver in response to short-term exposure to arsenite. Environmental and Molecular Mutagenesis. 2007;**48**(8):658-665

[97] Humtsoe N, Davoodi R, Kulkarni BG, Chavan B. Effect of arsenic on the enzymes of the Rohu carp, *Labeo rohita* (Hamilton 1822). The Raffles Bulletin of Zoology. 2007;**14**:17-19

[98] Patlolla AK, Tchounwou PB. Serum acetyl cholinesterase as a biomarker of arsenic induced neurotoxicity in Sprague-Dawley rats. International Journal of Environmental Research and Public Health. 2005;**2**(1):80-83

[99] Gül Ş, Belge-Kurutaş E, Yıldız E, Şahan A, Doran F. Pollution correlated modifications of liver antioxidant systems and histopathology of fish (Cyprinidae) living in Seyhan Dam Lake, Turkey. Environment International. 2004;**30**(5):605-609

[100] Pal S, Chatterjee AK. Protective effect of methionine supplementation on arsenic-induced alteration of glucose homeostasis. Food and Chemical Toxicology. 2004;**42**(5):737-742

Chapter 6

Exploring the Interplay between Arsenic and Cutaneous Physiology, Pathology, and Regeneration

Mangalathettu Binumon Thankachan, Gayathri S. Kamath, Greeshma Sasikumar and Sreejith Parameswara Panicker

Abstract

Arsenic poisoning and groundwater exposure are not regional hazards; we can call them a "silent global hazard." The victims are not always aware of arsenic-exposed daily life and the use of contaminated groundwater. The World Health Organization (WHO) reported that several countries, including Bangladesh, India, Argentina, Chile, Viet Nam, Cambodia, Pakistan, China, the United States of America (USA), and Mexico, have inorganic arsenic naturally present at high levels in the groundwater. Many of these countries exceeded the typical toxic risk index of arsenic level of the WHO standard of 10 μg L^{-1}. The skin is the primary barrier of the body, and compromising the function of the skin is the beginning of psychosocial and physiological discomfort in humans. Hair loss, skin pigmentation, and skin irritation are the leading psychosocial and physiological facts induced by exposure to arsenic contamination. Like hair, nails are susceptible to external harm from arsenic because they may absorb and accumulate arsenic in vitro. The normal architecture of the skin changes to form epithelial hyperplasia, epidermal erosion, hyperkeratosis, degeneration of skin glands, and gradual replacement of hair shaft to keratinized substance. The extreme condition of arsenic exposure ultimately result in various skin carcinomas and alopecia.

Keywords: fibrosis, melanoma, arsenic, keratosis, cancer, aging, skin

1. Introduction

Arsenic is an element abundantly distributed in the earth's crust. That way, arsenic was exposed to humans and other lives, causing a hazard in strongly occurring regions. Long known to be harmful to human health, arsenic is a common element in the earth's crust and is frequently found in small amounts in food, water, and air [1]. The use of lead arsenate and sodium arsenite as pesticides has since been phased out in the United States of America (USA), but monomethyl arsenate is still used in agriculture in significant quantities. High exposure to them once occurred from drugs such as Fowler's solution and chemical pesticides such as sodium arsenite and lead arsenate. Arsenic exposure can result from mining operations where workers breathe ore dust, smelting fumes, or those people who come into contact with the leaching from tailings into waterbodies. Creating glass, using high-arsenic coal to dry food, disposing of a lot of wood that has been chromated with copper arsenate, and using poultry litter that has been given organic arsenical antibiotics can all lead to high exposures. There are three different types of arsenic compounds: first, pentavalent, As^{+5} organic or arsenate compounds (such as alkyl arsenates); second, trivalent, As^{+3} inorganic or arsenate compounds (such as sodium arsenate, arsenic trioxide); and third, arsine gas, AsH3, an inert gas produced by the reaction of acids with arsenic. Compared to pentavalent As (As^{5+}), trivalent As (As^{3+}) is more soluble and mobile. Some microbes, such as some types of *Bacillus* and *Pseudomonas*, have the ability to oxidize or reduce arsenic [2]. The deadly and extremely volatile arsenes can be produced by *Penicillin brevicaule*, also known as the arsenic fungi. Fungi, yeasts, and bacteria can methylate arsenic to produce monomethyl arsonate, dimethylarginine, and gaseous variants of arsine, which are found in large quantities in soils [3]. According to [4], two rod-shaped gram-positive bacteria were to be able to eliminate about 50% of arsenite and arsenate from culture media containing arsenic that had been isolated from the Purbasthali block of West Bengal, India. It has been proven beyond a shadow of a doubt that one of the main contributors of arsenic in polluted groundwater throughout much of the world is the microbial mobilization of arsenic in the aqueous phase [5]. Understanding the processes used by purely cultured bacteria that were isolated from arsenic-polluted groundwater will deepen our understanding of the role of microbes in the biogeochemical cycle of arsenic. It may also make it easier to create effective management methods for groundwater resources that will provide affected populations with clean drinking water [6].

As a result of arsenic bacterial metabolisms shaping the lithosphere, arsenic biogeochemistry is inextricably linked to the evolution of microbes on Earth [7]. While extensive research has been done in recent years on the microbial communities that live in arsenic-affected Holocene and Pleistocene aquifers in Asia, little is known about the microbial communities responsible for the arsenic dissolution in polluted groundwater in Northern Italy [8].

2. Sources of arsenic exposure

Arsenic reaches out into the environment *via* various natural sources like volcanic eruptions. A broad, natural spreading occurs as a result of weathering and sedimentation. Arsenic is unintentionally released into the environment by fossil fuel combustion and non-ferrous metal smelting, typically as arsenic trioxide. In conjunction with sulfur, as in the majority of coals, arsenic builds up on fly ash particles during combustion together with other trace elements. Despite being present in very small amounts in coal, fly ash has a large amount of arsenic [9].

Arsenic enters the body in three ways: through the respiratory tract, through the gastrointestinal tract, and through the skin. The size and chemical makeup of the arsenic particles determine how far arsenic will penetrate into the lungs [10]. The amount of absorption in the gastrointestinal tract depends on the solubility of arsenic molecules [11]. Arsenic has demonstrated the capacity to disrupt the population of gut microbes, producing adverse responses in mice and people [12]. Around 95% of the dose consumed by soluble trivalent compounds of arsenic is absorbed from the gastrointestinal tract [13]. Acute gastrointestinal symptoms are more frequently observed following intake of arsenic than after inhalation or cutaneous absorption. The stomach and intestine can experience severe inflammation, necrosis, and perforation of the mucosa and submucosa. Bloody diarrhea may emerge as a symptom of hemorrhagic gastroenteritis [14]. Arsenic absorbed through the skin is poorly studied, and evidence suggests that arsenic trichloride and arsenic acid are absorbed *via* skin. The arsenic acid-induced dermal injury will promote absorption [10]. Arsenic exposure may occur while handling conserved wood items that contain it through direct skin contact. However, just barely is known about the chemical form, absorption circumstances, kinetics, or other details necessary to conclude skin absorption in certain populations [15]. Arsenic trichloride splashes on a worker's skin have been reported to have toxic consequences in the occupational literature [16].

Arsenic pollution is not a regional hazard because of the stratification of the compound naturally found throughout our environment. The World Health Organization (WHO) identified the world's major arsenic-exposed regions, including Bangladesh, India, Argentina, Chile, Viet Nam, Cambodia, Pakistan, China, the United States of America, and Mexico. The sample water from the most affected regions of Argentina has an approximately 88% of 86 groundwater sources exceeding the arsenic concentration according to the WHO guideline value [17] in 2007. The risk of arsenic poisoning from drinking water from tainted tube wells has been estimated to affect 50 million people in Bangladesh [18]. Arsenic groundwater pollution has been discovered in Cambodia, where around 100,000 family-based wells provide drinking water. A previous study of an extensive groundwater survey and seasonal fluctuations in the Mekong River floodplain comprised 3700 km^2 (131 samples, 30 parameters). The 1.2 million residents of this region face a health risk due to the arsenic levels, which range from 1 to 1340 gL^{-1}, with an average of 163 g L^{-1}, and 48% exceeding 10 g L^{-1}. The potential exposure of 350 people km^{-2} to chronic arsenic poisoning is comparable to Bangladesh's 200 km^{-2} magnitude [19]. The investigation of the long-term dangers of arsenic-caused cancer has numerous significant advantages due to an unusual arsenic exposure situation in northern Chile. River water from the nearby Andes Mountains that contained significant levels of naturally occurring arsenic was diverted to the area's major city (Antofagasta) for drinking in the late 1950s [20]. As a result, the city's water supply had an average arsenic concentration of 860 g/L for 13 years (1958 to 1970) [18, 21]. According to the West Bengal study's conclusions, arsenic-induced disease manifestation in humans may be caused by deficiencies in DNA repair ability, disruptions in the methylation of the p53 and p16 gene promoter regions, and changes in genomic methylation. Arsenic-induced keratosis has been found to be more common and associated with P53 polymorphism [22].

3. Impact of arsenic exposure: a global scenario

In the previous two to three centuries of the last completed millennium, the Bengal region faces the first incidence of arsenic detection in groundwater in India. The Ganga-Brahmaputra River delta, which creates the Bengal Basin, is where most of the arsenic in drinking water is found. This has been explained by the significant amounts of sediments rich in arsenic that these rivers carried downstream during the Pleistocene and Holocene epochs [23]. The detection of arsenic has been reported in the Indian states of Uttar Pradesh, Bihar, Assam, Chhattisgarh, Jharkhand, and Karnataka. Most of it is found in alluvial soil in Bihar, West Bengal, and Uttar Pradesh, whereas in Chhattisgarh, it is predominantly found in volcanic rocks. Now that populations in Bangladesh and India numbering in the tens of millions who are at risk make well water drawn from geological formations with high-arsenic levels a serious health problem, this issue is brought up [24]. In India, groundwater arsenic contamination and its health effects were documented for over 28 years (1988–2016) in the states of Jharkhand, West Bengal, Uttar Pradesh, and Bihar in the Ganga River floodplain and the states of Assam and Manipur in the Brahmaputra and Imphal river flood plains. Although Rajnandgaon village in Chhattisgarh state is not on a flood plain, the groundwater there is still contaminated with arsenic. After being analyzed, more than 170,000 samples of tube well water from the impacted states contained arsenic, with a maximum concentration of 3700 g/L in half of the samples [25].

Due to water contamination, several communities are at risk of a high prevalence of skin cancer, especially Chileans as well as some Taiwanese. Elderly people who exhibit indications of persistent arsenic poisoning are more likely to develop cancer. Animals generally have a much lower incidence of As-induced cancer [26]. Research conducted in the past on residents of southwest Taiwan revealed a connection between high-arsenic exposure and altered skin pigmentation, Blackfoot disease, hyperkeratosis of the palms and soles, which is brought on by poor circulation and endothelial cell damage, and skin cancer and several other organs such as the liver and lung bladder. Skin hyperpigmentation and keratosis occur at drinking water concentrations of around 50 parts per billion (ppb), the previous US drinking water norm, according to more recent concentration-dependent- tests conducted in Bangladesh [27]. India and Bangladesh have been claimed to be the most severely affected nations, with over 100 million people potentially at risk due to excessive levels of arsenic in groundwater [25].

It has been found that natural sources are the primary causes of this contamination. But only in a few zones have the precise mechanisms that release these toxic substances into groundwater been identified [28]. In many areas of Pakistan, arsenic is a significant problem, and contaminated groundwater causes substantial health risks. Lahore's rural areas have reported on early research on groundwater and the health dangers of arsenic poisoning, but no such studies have been carried out in the nation's major cities [29]. Aquifers in different parts of Mexico have been found to contain concentrations of arsenic and fluoride that are higher than those of Mexican drinking water standards.

Arsenic may travel long distances and adhere to tiny airborne particles, remaining in the atmosphere for several days. Mild exposure to arsenic may occur of its beneficial uses. Arsenic is frequently used as a pesticide, herbicide, or wood preservative due to its germicidal properties and the ability to withstand rot and decay [30]. Since many common compounds containing arsenic can dissolve in water, they can contaminate lakes, rivers, or underground water when they are exposed to rain, snow, or abandoned waste from industries. Because of this, arsenic pollution of groundwater poses a severe risk to human health globally.

4. Routes of arsenic exposure and its consecutive health hazards

Detrimental health consequences, such as hyperpigmentation, keratosis, different types of cancer, and vascular illnesses, have also been connected to long-term arsenic exposure from ingesting contaminated food or water or breathing contaminated air in a number of different nations [31, 32]. According to a recent article by the US Agency for Toxic Substances and Disease Registry, occupational workers are more likely to be exposed to arsenic through inhalation and the dermal layer than the general population.

There are no reliable studies to quantify the rate of arsenic absorption *via* healthy human skin. The skin may be a potential route of absorption in humans, according to occupational incidents where significant interaction of the skin with aqueous solutions of inorganic arsenic compounds resulted in systemic poisoning and increased urine arsenic excretion by the employees involved in their manufacture [33]. On the other hand, the spectrophotometric study reveals that mean arsenic levels in hair represent systemic absorption. When hair grows, it is absorbed into the matrix and only lost again when lost or removed [34, 35]. Arsenic is difficult to detect because it is found in hair in trace amounts, despite providing strong evidence in numerous investigations. The scalp is primarily affected by arsenic from food, work, and water [36].

Oral exposure to arsenic is the main way people get exposed to it. Thus, most diets are the main way people are exposed to arsenic. Routine activities may expose young children to small amounts of arsenic orally, which may be a significant exposure route. A pathologic sign of chronic exposure is patchy hyperpigmentation, which can occur anywhere on the body. And it occurs predominantly on the axillae, groin, neck, eyelids, temples, and nipples. "Raindrops on a dusty road" frequently describes the dark brown patches with scattered lighter specks. In extreme cases, the pigmentation covers a large portion of the chest, back, and belly. Pigment alterations have been seen in populations regularly consuming water with an arsenic content of 400 parts per billion (ppb) or more [14]. The palms and soles are the most common sites for arsenical hyperkeratosis. Topical application of arsenic for medical purposes will only contribute to total body load *via* direct dermal uptake of arsenic.

5. The fate of arsenic: a comprehensive overview of its biochemical aspect

After absorption, arsenic compounds typically undergo metabolic processing in the liver, where they are transformed into a variety of inorganic and organic species such as arsenite, dimethylarsinate, arsenate, and monomethylarsonate. Arsenic, both organic and inorganic, is quickly absorbed into the bloodstream and is circulated to the human digestive system. Since they are not well absorbed by tissues, organic arsenic forms are typically regarded as harmless [37]. Nevertheless, inorganic arsenic species are highly reactive and impact several intracellular processes [38]. Along with the significant role that methylated trivalent arsenic compounds play in the development of cancer, the ability of arsenic compounds to cause cancer in humans in the skin, lung, liver, and bladder appears to be best explained by the oxidative stress theory [24].

Arsenic exposure can come out of the environment *via* natural as well as man-made sources. Skin absorption is among the several ways through which arsenic enters the body [39, 40]. International Agency for Research on Cancer (IARC) has acknowledged arsenic as a class 1 carcinogen. Its effects were mediated through abnormal DNA repair and other cellular mechanisms [41, 42]. Arsenic has an impact on all organ systems in our body. Skin, the interface between the external and internal environments, is the most vulnerable organ in our body. Dermal impacts are the most sensitive and are considered the endpoint of arsenic exposure. Often, arsenic toxicity is manifested through the skin [43]. Often it takes about more than 20 years for its manifestation [44]. The time between exposure and the development of skin cancer can last up to 30 years. Long-term exposure to arsenic will lead to skin darkening and hyperkeratosis [45]. Dermal effects, such as melanosis, leukomelanosis, keratosis, Bowen's disease, and cancer, will result from chronic exposure to arsenic [46]. Pigment alterations and palmoplantar hyperkeratosis also characterize chronic arsenic exposure. Arsenical keratoses may develop into cancer. Mees lines in the nails may be a sign of the delayed consequences of acute or chronic exposure. Mees lines are horizontal ridges found in digit nails [14].

However, due to the local absorption of proteins containing sulfhydryl groups, the skin is a crucial organ for arsenic toxicity independent of the route of exposure. Despite the fact that arsenic poisoning's effects on the skin are characterized by multifocal lesions all over the body, chronic arsenic build-up increases the skin's vulnerability to ultraviolet (UV) radiation and is linked to an increased risk of tumors of exposed skin [47, 48]. Because of the high quantities of keratin in hair and skin, autoradiographic investigations on the concentrations of arsenic in animals reveal that these tissues have the highest levels of long-term retention [49]. Desquamation and hair loss are important methods of excreting arsenic due to its attraction for sulfhydryl groups, which causes build-up and persistent retention in keratin-rich tissues like hair, skin, and nails [50].

Liver damage, dermatological lesions, respiratory disorders, peripheral neuropathy, ocular inflammation, and irritations, make up the majority of the common health outcomes in humans to date. Recently, *in vitro* research using human cancer cell lines has revealed these harmful mechanisms linked to liver, neuronal, lungs, bladder, and dermatological health outcomes [51, 52]. With human cancer cell lines, studies on arsenic over the previous 10 years have demonstrated that arsenic alters the pattern of gene expression within the cell as well as epigenomic profile, telomere length, cell cycle, etc. [53, 54].

Hutchinson published the first account of arsenic-induced skin cancer and its connection to hyperkeratosis in patients exposed to medicinal arsenic in 1887 [55]. The human elementary canal is where arsenite chemical is mostly absorbed and deposits heavily in all kinds of bodily cells. As a result, it has an impact on the cell's enzyme function, and eventually the damaged cells progressively decompose [56].

The distinctive skin lesions included keratoses on the palms and soles and pigmentation changes, mostly on the upper chest, arms, and legs. Skin cancers, raindrop guttate hypopigmentation, excessive arsenical keratosis (scaly skin formation), excessive scaly skin formation on the palms and feet, and exfoliative acne are some of the other signs caused by arsenic (particularly Bowen's disease) [57]. The most prevalent skin alteration associated with chronic arsenic toxicity is hyperpigmentation [58].

People having age over 40 years had the highest prevalence of arsenical dermatosis. The prevalence did not differ by gender. A 60-year-old man from Murshidabad, West Bengal, showed several raised lesions on his palms and soles that had been present for 4 years and were painless and asymptomatic. Both palms displayed numerous hyperkeratotic papules upon cutaneous examination. The ulceroproliferative lesion's histopathological analysis suggested that it was a moderately differentiated squamous cell carcinoma. Hair and nail samples had substantial amounts of arsenic. He was diagnosed with arsenical keratoses and squamous cell carcinoma and underwent surgery and chemotherapy for his condition [59].

To comprehend the mechanism of toxicity and evaluate the health effects, biomarkers of inorganic arsenic exposure are required. The four biological substrates that are most frequently used in epidemiological studies are blood, urine, hair, and nails [60]. In addition to the key function that methylated trivalent arsenic compounds carry in cancer development, the oxidative stress theory appears to be the mechanism of action that best accounts for the ability of arsenic compounds to trigger cancer in humans in the skin, lung, bladder, and liver [24]. The presence of arsenic in these biological samples suggests that it was absorbed systemically after exposure. In addition, they can bind to hair and nails, which can serve as biomarkers of arsenic toxicity caused due to external exposure [61]. Despite the fact that human fingernails naturally contain arsenic, a study explored the possibility of using arsenic levels as a biological indicator of occupational exposure to this substance [62]. Arsenic uptake through drinking water ingestion is related to the concentration of arsenic in fingernails, a biomarker for human exposure.

In addition to the 40 healthy participants from the arsenic-affected rural areas of Iran, 49 fingernail samples were also collected from people who lived in areas where drinking water sources had not been reported to be contaminated with arsenic. It was found that the fingernail arsenic contents in 50 and 4.08% of the samples taken from arsenic-contaminated and reference villages were higher than the typical arsenic values of nails (0.43–1.08 g/g) [63]. The measurement of total arsenic in nails can be performed using the inductively coupled plasma-mass spectrometry (ICP-MS) technique [64].

It can be challenging to select an appropriate biomarker to study arsenic exposure. A biomarker of arsenic exposure from drinking water is the total arsenic concentration in blood or urine [65]. Contact with water and particles containing arsenic can contaminate nails, hair, and skin, but chemical hair therapies can change the rate at which arsenic accumulates in the body [66].

6. Long-term arsenic exposure: chronic health impacts

Drinking water contaminated with arsenic can have long-lasting negative effects on the condition of one's physical well-being. The prevalence of skin lesions and arsenic concentration in drinking water have a dose-response relationship, according to sizable population research conducted in West Bengal, India [67].

Consumption of water containing arsenic can lead to adverse health effects such as cutaneous abnormalities and lesions. Keratosis and pigment disorders, including hyperpigmentation and hypopigmentation, are the diagnoses. According to a particular study, compared to the exposed group, the incidence of hyperkeratosis, including Palmar hyperkeratosis, was 34 times higher than in the control group [68]. The most recent estimate of the population exposed to arsenic concentrations over the WHO safety threshold (10 g/L) for drinking water showed that about 140 million people in atleast 70 countries have been affected. These findings imply that, despite the most recent recommendations for the maximum allowable limit of arsenic in drinking water, the population that has been chronically exposed to arsenic for a long time may still be in danger.

The majority of earlier studies on arsenic toxicity focused on population-based epidemiological outcomes, analyses of particular disease risks, chemical-based and physiologic aspects of arsenic metabolic processes, and research on related gene expression profiles, DNA damage, and cancer; all of which were linked to the process of toxicity and subsequent consequence or manifestation of disease. To explain differences in susceptibility to arsenic exposure, research is currently shifting toward the study of epigenetic modifications (DNA methylation, microRNA (miRNA), and histone modification) [69].

6.1 Exploring the link between arsenic exposure and skin lesions

One of the most prevalent signs of chronic arsenic poisoning is skin lesions. A rise in the frequency of skin lesions was found even at a mild exposure range of 0.005–0.01 mg/l of arsenic in drinking water [70].

Arsenical skin lesions were recognized when at least one of the following criteria was met: changes in the body's covered surfaces' pigmentation and/or keratosis, especially on the palms and toes. Anywhere on the body, melanosis can appear, and its typical symptoms include raindrop-like pigmentation or extensive dark brown dappling in non-exposed areas of the body. The distribution of leukomelanosis is the same and it can exist even when there is no melanosis. Keratosis is characterized by small, nodular elevations resembling corns, typically 0.4–1 cm in diameter. It can be seen on the lateral edges of the palms, soles, fingers, heels, and toes. On the soles and palms, diffuse keratosis can also be seen. Melanosis and keratosis are the most obvious skin lesions linked to arsenic toxicity, which can eventually turn into skin cancers. There were 675 subjects with keratosis and 1135 subjects with melanosis overall in the survey that was conducted in Bangladesh, with a sex ratio of 1.5:1 (men:women) [71].

In addition to accumulating other trace elements, toes can also accumulate arsenic. This suggests that there are frequent exposure sources and metabolic pathways [72]. Toenail arsenic is also linked to cadmium, manganese, and lead concentrations [73]. According to research, the total arsenic concentrations in toenails can serve as a biomarker for prolonged exposure. They are being studied because they can accumulate arsenic, have a slow growth rate, and may be less susceptible to external contamination than samples from fingernails and hair [74]. Arsenic content in toenails and several cancers is correlated, including squamous cell skin cancer, which has been found [75]. Higher toenail arsenic concentrations were linked to an increased risk of keratosis in Bangladesh and India [76]. Due to arsenic's high affinity for keratin, hair contains more of it than other tissues. In an examination of samples taken from Chilean villagers, inorganic arsenic, commonly known as inorganic arsenic, was the most common kind of arsenic found in human hair [77].

6.2 The cause and devastating impact of arsenic-induced skin cancer

Skin cancer and other internal cancers are known to be caused by the known cancer-causing agent—arsenic. Cancer is slowly developed after decades (approximately 20 years) after exposure to contact with the carcinogen [78]. The abnormal epidermal keratinocyte growth, differentiation, dysplasia, and dermal inflammatory infiltrates that are the hallmark of arsenical skin cancer may be due to mitochondrial control of cell proliferation, energy production, reactive oxygen species (ROS) production, DNA damage and mutations, and immune control [41].

These dermatological lesions and the risk of skin cancer are significantly correlated. Basal cell carcinoma (BCC), Bowen's disease, and squamous cell carcinoma (SCC) are the three skin cancers most frequently triggered once exposed to arsenic [45]. Strong evidence suggests that only highly exposed populations exhibit a linear dose-response relationship between the concentration of arsenic and cancer risks. The dose-response curve may be influenced by several factors, including gender, ionizing radiation, smoking, diet, and genetic susceptibility, which may act synergistically or as confounders [24].

In the region of Taiwan where persistent arsenic breakouts take place, a study was conducted on the prevalence of skin cancer. The majority of the population had

Blackfoot disease, which showed up as keratosis and hyperpigmentation on the palms and soles. According to a survey on chronic arsenicism conducted on 40,421 people in 37 villages, 360 had Blackfoot disease, 428 had skin cancer, 7418 had hyperpigmentation, and 2868 had keratosis. In these circumstances, they commonly coexisted with one another. The most prevalent types of skin cancer, epidermoid or basal cell carcinomas, typically appear on exposed parts including the head, face, and extremities [79].

6.3 Arsenic's societal impact: a growing cause for concern

The socioeconomic and demographic circumstances of the population exposed to arsenic dangers worsen its effects on human health [80]. Typically, underprivileged people suffer the most and are most at risk when arsenic levels in food and drinking water are high [81]. Children are the most susceptible age group and are at risk for the build-up of heavy metals since they require more energy and water per body weight than adults [82]. Arsenic at greater concentrations can result in various acute arsenic poisoning symptoms, including vomiting, abdominal discomfort, and diarrhea. The following symptoms are sensations of numbness and tingling in the extremities, cramping muscles, and in severe cases, death [83].

Numerous studies have examined the association between long-term exposure to arsenic and multiple medical conditions, including Blackfoot disease, cardiovascular and cerebrovascular diseases, chromosomal abnormalities, diabetes, hypertension, and peripheral vascular disease [68]. Arsenic readily crosses the blood-brain barrier and can build up in the striatum and hippocampus, among other brain regions. This increases arsenic toxicity and tissue damage [84]. Numerous neurological disorders are known to be brought on by arsenic exposure through various molecular mechanisms, including cytotoxicity, cellular DNA damage, chromosomal abnormalities, and a rise in the production of reactive oxygen species [85]. Capillary damage and dilatation also occur, causing fluid to transude, which in turn reduces blood volume and results in circulatory collapse. Blackfoot is a prevalent disease in Taiwan that is brought on by arsenic (As) and is characterized by the loss of blood flow to the extremities, which causes gangrene [26]. Arsenic-induced capillary alterations bring on kidney tubular degeneration.

6.4 What role does arsenic play in signaling pathways and stem cell functioning?

Signaling pathways can be impacted by arsenic. For instance, exposure to arsenic can result in neuronal cell death *via* activation of the nuclear factor-kappa B (NF-kB) and mitogen-activated protein kinase (MAPK) pathways [86, 87]. Arsenite (3 μM) exposure activates the serine-threonine liver kinase B1-AMP-activated protein kinase (LKB1-AMPK) signaling pathway and can suppress neuronal expansion [88]. It has the ability to modulate the protein kinase B (AKT) pathway to prevent myocyte differentiation and muscle regeneration [89]. Low amounts of arsenic seem to have the ability to inhibit stem cell development in stem cells *via* the Wnt/β-catenin pathway. The viral skin pathogen can interact with the effect of arsenic on skin. Arsenic exposures reduce the immunological response, at least in part, by reducing dendritic cell immune surveillance and cluster of differentiation 4 (CD4) cell activation [90]. Arsenic weakens the immune system's ability to fight off infections. Arsenic exposure has been shown to have this effect for influenza A, which raises viral titers and morbidity [91]. Similar to arsenicosis, human papillomavirus (HPV), a pathogen that only affects humans, shares several clinical traits with it, and may be a factor in arsenical skin disease. By avoiding recognition by Langerhans cells, cutaneous HPV develops infection. Therefore, it makes sense that arsenic's immune suppression could expose preexisting illnesses or weaken the immune system's response to fresh exposures [92]. In addition to its impact on the functioning of the immune system, arsenic may facilitate the process of HPV-mediated neoplasia, which involves the incorporation of HPV DNA into the keratinocyte genome. The expression of genes that encourage keratinocyte proliferation and hinder differentiation is brought on by damage to the episomal HPV DNA, such as that forced on by oxidative stress [93]. Interleukin-1 (IL-1) production in the murine keratinocyte cell line HEL30, which is linked to skin cancer, has been greatly attributed to arsenic [94]. There have been studies on arsenic-induced myocardial aberrations. When applied to rats, arsenic decreased cardiomyocyte viability, increased ROS production, and caused apoptotic cell death by activating caspase-3 and cleaving poly-ADP ribose polymerase (PARP). This increased IkB kinases (IKK) and NF-kB (p65) phosphorylation via an oxidative-mediated pathway, and ultimately caused cardiac apoptosis [95]. A transfected thyroid hormone receptor-mediated gene response element-luciferase construct and the endogenous thyroid receptor-regulated type I deiodinase (DIO1) gene were both significantly altered by treatment of human embryonic NT2 with low concentrations of arsenite (0.01–5 μM sodium arsenite) [96].

6.5 Aging

The epigenome may play a role in the health effects of arsenic and act as a biomarker for exposure. Numerous epigenetic biomarkers have been developed to measure various aspects of aging, including chronological age, morbidity, and mortality, in particular tissue types. Ongoing research on the health effects of aging as in Bangladesh revealed that middle-aged men had the highest incidence of as-induced skin lesions [97]. Organ pathophysiology is directly correlated with age, and oxidative stress negatively impacts the body and worsens with age [98]. There is a correlation between epigenetic age acceleration and prenatal and early-life arsenic exposure [99]. The balance between adipogenic and osteogenic differentiation may be impacted by arsenic exposure's promoting senescence in human mesenchymal stem cells. In a study conducted comparing women without arsenic skin lesions, those with arsenic skin lesions were 1.5 years younger at the time of menopause [100].

6.6 Strategies for the management of arsenicosis

Arsenicosis has no known treatment, so the best course of action is to stop drinking water that has been contaminated with arsenic [101]. The availability, effectiveness, and development potential of nearby alternative water sources will determine the region's choice of water source. In most areas with rainfall, safe surface water for drinking, cooking, or collecting rainwater has good potential. It can be used in conjunction with technology found in the average home if sufficient storage tanks are provided. This approach is especially helpful where there are few sources of surface water in sufficient quantities and of high quality. For the Treatment of arsenic-contaminated water, various options are available depending on technologies, cost, and acceptability and range from filter units for domestic use, through filter units for communitylevel use to piped supply of arsenic-treated water. Arsenic-contaminated water can be treated using a range of methods. Filter units for domestic use, community use, and piped supplies of water treated for arsenic are all options, depending on technologies, cost, and acceptability. Studies have shown that medications, various types of herbal remedies, vitamin C supplements, and other dietary supplements may reduce the effects of arsenic toxicity [102].

- Avoiding drinking water and other sources that may have high levels of arsenic.
- Setting up a mechanism to monitor drinking water quality and developing inter-sectoral connections to ensure a steady supply of arsenic-free drinking water are both essential.
- Treatment is the use of nutrition and medicines to hasten recovery and prevent further sickness.
- Arsenic in urine, hair, and nails can be tested for by establishing ties to existing diagnostic centers.
- The administration of all-encompassing palliative care in an effort to resolve specific problems or alleviate bodily symptoms.
- Medical observation is a secondary method for avoiding undesired outcomes.
- Offering sufficient therapy, education, and rehabilitation to deal with the psychological fallout of the disease.

7. Conclusion

Arsenic is present in small amounts in food, water, and the air in the earth's crust. By burning fossil fuels and smelting non-ferrous metals, arsenic is unintentionally released into the environment, usually as arsenic trioxide. Arsenic enters the body in three different ways: through the skin, through the gastrointestinal tract, and through the respiratory tract.

Arsenic compounds are usually metabolically processed in the liver following absorption, where they are converted into various inorganic and organic species such as arsenite, dimethylarsinate, arsenate, and monomethylarsonate. The major areas of the world exposed to arsenic include Bangladesh, India, Argentina, Chile, Viet Nam, Cambodia, Pakistan, China, the United States of America (USA), and Mexico, according to the World Health Organization (WHO). Fifty million people may be at risk of arsenic poisoning after consuming water from contaminated tube wells. Skin lesions are one of the most common symptoms of chronic arsenic poisoning. In several nations, long-term arsenic exposure from consuming contaminated food or water or breathing contaminated air has also been linked to detrimental health effects such as hyperpigmentation, keratosis, various types of cancer, and vascular illnesses.

Indicators of arsenic toxicity brought on by external exposure include hair and nails. Many epigenetic biomarkers have been developed to measure different aspects of aging, such as chronological age, morbidity, and mortality, in specific tissue types. To prevent the effects of arsenic toxicity, it is best to stop drinking water that has been contaminated with arsenic and start taking dietary supplements.

Author details

Mangalathettu Binumon Thankachan, Gayathri S. Kamath, Greeshma Sasikumar and Sreejith Parameswara Panicker*
Department of Zoology, University of Kerala, Thiruvananthapuram, India

*Address all correspondence to: p.sreejith@gmail.com; psreejith@keralauniversity.ac.in

References

[1] Hughes MF, Beck BD, Chen Y, Lewis AS, Thomas DJ. Arsenic exposure and toxicology: A historical perspective. Toxicological Sciences. 2011;**123**(2):305-332

[2] Yan-Chu H. Arsenic distribution in soils. In: Nriagu JO, editor. Arsenic in the Environment. Part I: Cycling and Characterization. New York: Wiley; 1994. pp. 17-50

[3] Takamatsu T, Aoki H, Yoshida T. Determination of arsenate, arsenite, monomethylarsonate, and dimethylarsinate in soil polluted with arsenic. Soil Science. 1982;**133**(4):239-246

[4] Dey U, Chatterjee S, Mondal NK. Isolation and characterization of arsenic-resistant bacteria and possible application in bioremediation. Biotechnology reports. 2016;**10**:1-7

[5] DrewniakL,MaryanN,LewandowskiW, Kaczanowski S, Sklodowska A. The contribution of microbial mats to the arsenic geochemistry of an ancient gold mine. Environmental Pollution. 2012;**162**:190-201

[6] Sarkar A, Kazy SK, Sar P. Characterization of arsenic resistant bacteria from arsenic rich groundwater of West Bengal, India. Ecotoxicology. 2013;**22**:363-376

[7] Zhu YG, Yoshinaga M, Zhao FJ, Rosen BP. Earth abides arsenic biotransformations. Annual Review of Earth and Planetary Sciences. 2014;**42**:443-467

[8] Cavalca L, Zecchin S, Zaccheo P, Abbas B, Rotiroti M, Bonomi T, et al. Exploring biodiversity and arsenic metabolism of microbiota inhabiting arsenic-rich groundwaters in northern Italy. Frontiers in Microbiology. 2019;**10**:1480

[9] Piver WT. Mobilization of arsenic by natural and industrial processes. In: Fowler BA, editor. Biological and Environmental Effects of Arsenic. Amsterdam: Elsevier; 1983;**6**(1)

[10] Vahter MARIE. Metabolism of arsenic. Biological and Environmental Effects of Arsenic. 1983;**6**:171-198

[11] Tam GKH, Charbonneau SM, Bryce F, Pomroy C, Sandi E. Metabolism of inorganic arsenic (74As) in humans following oral ingestion. Toxicology and Applied Pharmacology. 1979;**50**(2):319-322

[12] Choiniere J, Wang L. Exposure to inorganic arsenic can lead to gut microbe perturbations and hepatocellularcarcinoma. Acta Pharmaceutica Sinica B. 2016;**6**(5):426-429

[13] Klein CB, Leszczynska J, Hickey C, Rossman TG. Further evidence against a direct genotoxic mode of action for arsenic-induced cancer. Toxicology and Applied Pharmacology. 2007;**222**(3):289-297

[14] Agency for Toxic Substances and Disease Registry (ATSDR). Toxicological profile for Arsenic. Atlanta, GA: U.S. Department of Health and Human Services, Public Health Service; 2007

[15] National Research Council [US— Committee on Medical and Biological Effects of Environmental Pollutants]. Arsenic: medical and biological effects of environmental pollutants. In: Arsenic: medical and biologic effects of environmental pollutants. Washington: National Academic Press; 1977

[16] Garb LG, Hine CH. Arsenical neuropathy: Residual effects following

acute industrial exposure. Journal of Occupational Medicine: Official Publication of the Industrial Medical Association. 1977;**19**(8):567-568

[17] Litter MI, Ingallinella AM, Olmos V, Savio M, Difeo G, Botto L, et al. Arsenic in Argentina: Occurrence, human health, legislation and determination. Science of the Total Environment. 2019;**676**:756-766

[18] Ahmad SA, Khan MH, Haque M. Arsenic contamination in groundwater in Bangladesh: Implications and challenges for healthcare policy. Risk Management and Healthcare Policy. 2018;**11**:261

[19] Buschmann J, Berg M, Stengel C, Sampson ML. Arsenic and manganese contamination of drinking water resources in Cambodia: Coincidence of risk areas with low relief topography. Environmental Science & Technology. 2007;**41**:2146-2152

[20] Fraser B. Cancer cluster in Chile linked to arsenic contamination. Lancet. 2012;**379**:603

[21] Steinmaus CM, Ferreccio C, Romo JA, Yuan Y, Cortes S, Marshall G, et al. Drinking water arsenic in northern Chile: High cancer risks 40 years after exposure cessation. Cancer Epidemiology, Biomarkers & Prevention. 2013;**22**:623-630

[22] Mazumder DG, Dasgupta UB. Chronic arsenic toxicity: Studies in West Bengal, India. The Kaohsiung Journal of Medical Sciences. 2011;**27**(9):360-370

[23] Tiwari R, Satwik S, Khare P, Rai S. Arsenic contamination in India: Causes, effects and treatment methods. International Journal of Engineering, Science and Technology. 2021;**13**(1):146-152

[24] Tapio S, Grosche B. Arsenic in the aetiology of cancer. Mutation Research/Reviews in Mutation Research. 2006;**612**(3):215-246

[25] Chakraborti D, Rahman MM, Das B, Chatterjee A, Das D, Nayak B, et al. Groundwater arsenic contamination and its health effects in India. Hydrogeology Journal. 2017;**25**:1165-1181

[26] Eisler R. A review of arsenic hazards to plants and animals with emphasis on fishery and wildlife resources. Advances in Environmental Science and Technology-New York. 1994;**27**:185-185

[27] Argos M, Kalra T, Pierce BL, Chen Y, Parvez F, Islam T, et al. A prospective study of arsenic exposure from drinking water and incidence of skin lesions in Bangladesh. American Journal of Epidemiology. 2011;**174**(2):185-194

[28] Armienta MA, Segovia N. Arsenic and fluoride in the groundwater of Mexico. Environmental Geochemistry and Health. 2008;**30**:345-353

[29] Sultana J, Farooqi A, Ali U. Arsenic concentration variability, health risk assessment, and source identification using multivariate analysis in selected villages of public water system, Lahore, Pakistan. Environmental Monitoring and Assessment. 2014;**186**:1241-1251

[30] Nriagu JO. Arsenic in the Environment. New York: Wiley; 1994

[31] Argos M, Kalra T, Rathouz PJ, Chen Y, Pierce B, Parvez F, et al. Arsenic exposure from drinking water, and all-cause and chronic-disease mortalities in Bangladesh (HEALS): A prospective cohort study. Lancet (London, England). 2010;**376**:252-258

[32] Gibb H, Haver C, Gaylor D, Ramasamy S, Lee JS, Lobdell D, et al. Utility of recent studies to assess the National Research Council 2001

estimates of cancer risk from ingested arsenic. Environmental Health Perspectives. 2011;**119**:284-290

[33] Buchanan WD. Toxicity of arsenic compounds. Toxicity of Arsenic Compounds. New York: Elsevier Publishing Company; 1962

[34] Chatt A, Secord CA, Tiefenbach B, Jervis RE. Scalp hair as a monitor of community exposure to environmental pollutants. In: Hair, Trace Elements, and Human Illness. New York: Praeger Publishers; 1980

[35] Hammer DI, Finklea JF, Hendricks RH, Shy CM, Horton RJM. Hair trace metal levels and environmental exposure. American Journal of Epidemiology. 1971;**93**(2):84-92

[36] Rasheed H, Kay P, Slack R, Gong YY, Carter A. Human exposure assessment of different arsenic species in household water sources in a high risk arsenic area. Science of the Total Environment. 2017;**584**:631-641

[37] Akter KF, Owens G, Davey DE, Naidu R. Arsenic speciation and toxicity in biological systems. Reviews of Environmental Contamination and Toxicology. 2005;**184**:97-149

[38] Drobná Z, Walton FS, Paul DS, Xing W, Thomas DJ, Stýblo M. Metabolism of arsenic in human liver: The role of membrane transporters. Archives of Toxicology. 2010;**84**:3-16

[39] Abdul KSM, Jayasinghe SS, Chandana EP, Jayasumana C, De Silva PMC. Arsenic and human health effects: A review. Environmental Toxicology and Pharmacology. 2015;**40**(3):828-846

[40] Sawada N. Association between arsenic intake and cancer-from the viewpoint of epidemiological study. Nihon eiseigakuzasshi. Japanese Journal of Hygiene. 2018;**73**(3):265-268

[41] Huang HW, Lee CH, Yu HS. Arsenic-induced carcinogenesis and immune dysregulation. International Journal of Environmental Research and Public Health. 2019;**16**(15):2746

[42] Zhou Q, Xi S. A review on arsenic carcinogenesis: Epidemiology, metabolism, genotoxicity and epigenetic changes. Regulatory Toxicology and Pharmacology. 2018;**99**:78-88

[43] Rahman MM, Ng JC, Naidu R. Chronic exposure of arsenic via drinking water and its adverse health impacts on humans. Environmental Geochemistry and Health. 2009;**31**:189-200

[44] Ghosh N, Kalman D, Smith MM, Mitra S, Santra A, Lahiri S, et al. Arsenic in drinking water and skin lesions: dose-response data from West Bengal, India. Epidemiology (Cambridge, Mass). 2003;**14**(2):174-182. DOI: 10.1097/01.EDE.0000040361.55051.54

[45] Yu HS, Liao WT, Chai CY. Arsenic carcinogenesis in the skin. Journal of Biomedical Science. 2006;**13**(5):657-666

[46] National Research Council (NRC). Arsenic in drinking water—2001 update. Washington, DC: National Academy Press; 2001

[47] Lüchtrath H. The consequences of chronic arsenic poisoning among Moselle wine growers: Pathoanatomical investigations of post-mortem examinations performed between 1960 and 1977. Journal of Cancer Research and Clinical Oncology. 1983;**105**:173-182

[48] Zaldívar R, Prunés L, Ghai GL. Arsenic dose in patients with cutaneous carcinomata and hepatic

haemangio-endothelioma after environmental and occupational exposure. Archives of Toxicology. 1981;**47**:145-154

[49] Lindgren A, Vahter M, Dencker L. Autoradiographic studies on the distribution of arsenic in mice and hamsters administered 74As-arsenite or-arsenate. Acta Pharmacologica et Toxicologica. 1982;**51**(3):253-265

[50] Molin L, Wester PO. The estimated daily loss of trace elements from normal skin by desquamation. Scandinavian Journal of Clinical and Laboratory Investigation. 1976;**36**(7):679-682

[51] Alarifi S, Ali D, Alkahtani S, Siddiqui MA, Ali BA. Arsenic trioxide-mediated oxidative stress and genotoxicity in human hepatocellular carcinoma cells. Oncotargets and Therapy. 2013;**6**:75-84. DOI: 10.2147/OTT.S38227

[52] Kundu M, Ghosh P, Mitra S, Das JK, Sau TJ, Banerjee S, et al. Precancerous and non-cancer disease endpoints of chronic arsenic exposure: The level of chromosomal damage and XRCC3 T241M polymorphism. Mutation Research/Fundamental and Molecular Mechanisms of Mutagenesis. 2011;**706**(1-2):7-12

[53] Calatayud M, Devesa V, Vélez D. Differential toxicity and gene expression in Caco-2 cells exposed to arsenic species. Toxicology Letters. 2013;**218**(1):70-80

[54] Li H, Engström K, Vahter M, Broberg K. Arsenic exposure through drinking water is associated with longer telomeres in peripheral blood. Chemical Research in Toxicology. 2012;**25**(11):2333-2339

[55] Neubauer O. Arsenical cancer; a review. British Journal of Cancer. 1947;**1**(2):192-251

[56] Saha JC, Dikshit AK, Bandyopadhyay M, Saha KC. A review of arsenic poisoning and its effects on human health. Critical Reviews in Environmental Science and Technology. 1999;**29**(3):281-313

[57] Smith AH, Lingas EO, Rahman M. Contamination of drinking-water by arsenic in Bangladesh: A public health emergency. Bulletin of the World Health Organization. 2000;**78**(9):1093-1103

[58] Yue-zhen H, Xu-chun Q, Guo-quan W, et al. Endemic chronic arsenism in Xinjiang. Chinese Medical Journal. 1985;**98**:219-222

[59] Neema S, Radhakrishnan S. Arsenical keratoses with squamous cell carcinoma. Indian Journal of Dermatology, Venereology and Leprology. 2022;**88**:376-377

[60] Marchiset-Ferlay N, Savanovitch C, Sauvant-Rochat MP. What is the best biomarker to assess arsenic exposure via drinking water? Environment International. 2012;**39**(1):150-171

[61] Hindmarsh JT. Caveats in hair analysis in chronic arsenic poisoning. Clinical Biochemistry. 2002;**35**(1):1-11

[62] Agahian B, Lee JS, Nelson JH, Johns RE. Arsenic levels in fingernails as a biological indicator of exposure to arsenic. American Industrial Hygiene Association Journal. 1990;**51**(12):646-651

[63] Shokoohi R, Khazaei M, Karami M, et al. Application of fingernail samples as a biomarker for human exposure to arsenic-contaminated drinking waters. Scientific Reports. 2022;**12**:4733

[64] Chen KLB, Amarasiriwardena CJ, Christiani DC. Determination of total arsenic concentrations in nails by inductively coupled plasma mass

spectrometry. Biological Trace Element Research. 1999;**67**:109-125

[65] Mandal BK, Ogra Y, Suzuki KT. Speciation of arsenic in human nail and hair from arsenic-affected area by HPLC-inductively coupled argon plasma mass spectrometry. Toxicology and Applied Pharmacology. 2003;**189**(2):73-83

[66] Hinwood A, Horwitz P, Rogan R. Human exposure to metals in groundwater affected by acid sulfate soil disturbance. Archives of Environmental Contamination and Toxicology. 2008;**55**:538-545

[67] Mazumder DNG, Haque R, Ghosh N, De BK, Santra A, Chakraborty D, et al. Arsenic levels in drinking water and the prevalence of skin lesions in West Bengal, India. International Journal of Epidemiology. 1998;**27**(5):871-877

[68] Dastgiri S, Mosaferi M, Fizi MA, Olfati N, Zolali S, Pouladi N, et al. Arsenic exposure, dermatological lesions, hypertension, and chromosomal abnormalities among people in a rural community of Northwest Iran. Journal of Health, Population, and Nutrition. 2010;**28**(1):14

[69] Sanyal T, Bhattacharjee P, Paul S, Bhattacharjee P. Recent advances in arsenic research: Significance of differential susceptibility and sustainable strategies for mitigation. Frontiers in Public Health. 2020;**8**:464

[70] Yoshida T, Yamauchi H, Sun GF. Chronic health effects in people exposed to arsenic via the drinking water: Dose-response relationships in review. Toxicology and Applied Pharmacology. 2004;**198**(3):243-252

[71] Milton AH, Rahman M. Environmental pollution and skin involvement pattern of chronic arsenicosis in Bangladesh. Journal of Occupational Health. 1999;**41**(4):207-208

[72] Signes-Pastor AJ, Carey M, Vioque J, Navarrete-Muñoz E, Rodríguez-Dehli C, Tardón A, et al. Urinary arsenic speciation in children and pregnant women from Spain. Exposure and Health. 2017;**9**:105-111

[73] Mordukhovich I, Wright RO, Hu H, Amarasiriwardena C, Baccarelli A, Litonjua A, et al. Associations of toenail arsenic, cadmium, mercury, manganese, and lead with blood pressure in the normative aging study. Environmental Health Perspectives. 2012;**120**(1):98-104

[74] Signes-Pastor AJ, Gutiérrez-González E, García-Villarino M, Rodríguez-Cabrera FD, López-Moreno JJ, Varea-Jiménez E, et al. Toenails as a biomarker of exposure to arsenic: A review. Environmental Research. 2021;**195**:110286

[75] Karagas MR, Stukel TA, Morris JS, Tosteson TD, Weiss JE, Spencer SK, et al. Skin cancer risk in relation to toenail arsenic concentrations in a US population-based case-control study. American Journal of Epidemiology. 2001;**153**(6):559-565

[76] Maity JP, Nath B, Kar S, Chen CY, Banerjee S, Jean JS, et al. Arsenic-induced health crisis in peri-urban Moyna and Ardebok villages, West Bengal, India: An exposure assessment study. Environmental Geochemistry and Health. 2012;**34**:563-574

[77] Katz SA. On the use of hair analysis for assessing arsenic intoxication. International Journal of Environmental Research and Public Health. 2019;**16**(6):977

[78] Jonnalagadda SB, Rao PP. Toxicity, bioavailability and metal speciation.

Comparative Biochemistry and Physiology Part C: Pharmacology, Toxicology and Endocrinology. 1993;**106**(3):585-595

[79] Tseng WP, Chu H, How SW, Fong JM, Lin CS, Yeh SHU. Prevalence of skin cancer in an endemic area of chronic arsenicism in Taiwan. Journal of the National Cancer Institute. 1968;**40**(3):453-463

[80] Singh SK, Vedwan N. Mapping composite vulnerability to groundwater arsenic contamination: An analytical framework and a case study in India. Natural Hazards. 2015;**75**:1883-1908

[81] Curry A, Carrin G, Bartram J, Yamamura S, Sims J, Hueb J, et al. Towards an Assessment of the Socioeconomic Impact of Arsenic Poisoning in Bangladesh (No. WHO/SDE/WSH/00.4). World Health Organization; 2000

[82] Schrey P, Wittsiepe J, Budde U, Heinzow B, Idel H, Wilhelm M. Dietary intake of lead, cadmium, copper and zinc by children from the German North Sea islandAmrum. International Journal of Hygiene and Environmental Health. 2000;**203**(1):1-9

[83] Flanagan SV, Johnston RB, Zheng Y. Arsenic in tube well water in Bangladesh: Health and economic impacts and implications for arsenic mitigation. Bulletin of the World Health Organization. 2012;**90**:839-846

[84] Hong YS, Song KH, Chung JY. Health effects of chronic arsenic exposure. Journal of Preventive Medicine and Public Health. 2014;**47**(5):245

[85] Wang Y, Zhao F, Liao Y, Jin Y, Sun G. Effects of arsenite in astrocytes on neuronal signaling transduction. Toxicology. 2013;**303**:43-53

[86] Felix K, Manna SK, Wise K, Barr J, Ramesh GT. Low levels of arsenite activates nuclear factor-κB and activator protein-1 in immortalized mesencephalic cells. Journal of Biochemical and Molecular Toxicology. 2005;**19**(2):67-77

[87] Yen CC, Ho TJ, ChoiniereWu CC, Chang CF, Su CC, Chen YW, el al. . Inorganic arsenic causes cell apoptosis in mouse cerebrum through an oxidative stress-regulated signaling pathway. Archives of Toxicology. 2011;**85**:565-575

[88] Wang X, Meng D, Chang Q, Pan J, Zhang Z, Chen G, et al. Arsenic inhibits neurite outgrowth by inhibiting the LKB1–AMPK signaling pathway. Environmental Health Perspectives. 2010;**118**(5):627-634

[89] Yen YP, Tsai KS, Chen YW, Huang CF, Yang RS, Liu SH. Arsenic inhibits myogenic differentiation and muscle regeneration. Environmental Health Perspectives. 2010;**118**(7):949-956

[90] Liao WT, Yu CL, Lan CCE, Lee CH, Chang CH, Chang LW, et al. Differential effects of arsenic on cutaneous and systemic immunity: Focusing on CD4+ cell apoptosis in patients with arsenic-induced Bowen's disease. Carcinogenesis. 2009;**30**(6):1064-1072

[91] Kozul CD, Ely KH, Enelow RI, Hamilton JW. Low-dose arsenic compromises the immune response to influenza a infection in vivo. Environmental Health Perspectives. 2009;**117**(9):1441-1447

[92] Frazer IH, Thomas R, Zhou J, Leggatt GR, Dunn L, McMillan N, et al. Potential strategies utilised by papillomavirus to evade host immunity. Immunological Reviews. 1999;**168**(1):131-142

[93] Jones EE, Wells SI. Cervical cancer and human papillomaviruses: Inactivation of retinoblastoma and other tumor suppressor pathways. Current Molecular Medicine. 2006;**6**(7):795-808

[94] Banerjee S, Datta S, Chattyopadhyay D, Sarkar P. Arsenic accumulating and transforming bacteria isolated from contaminated soil for potential use in bioremediation. Journal of Environmental Science and Health, Part A. 2011;**46**(14):1736-1747

[95] Sankar P, Telang AG, Kalaivanan R, Karunakaran V, Suresh S, Kesavan M. Oral nanoparticulate curcumin combating arsenic-induced oxidative damage in kidney and brain of rats. Toxicology and Industrial Health. 2016;**32**(3):410-421

[96] Davey JC, Nomikos AP, Wungjiranirun M, Sherman JR, Ingram L, Batki C, et al. Arsenic as an endocrine disruptor: Arsenic disrupts retinoic acid receptor–and thyroid hormone receptor–mediated gene regulation and thyroid hormone–mediated amphibian tail metamorphosis. Environmental Health Perspectives. 2008;**116**(2):165-172

[97] Rahman MS, Reichelt-Brushet AJ, Clark MW, Farzana T, Yee LH. Arsenic bio-accessibility and bioaccumulation in aged pesticide contaminated soils: A multiline investigation to understand environmental risk. Science of the Total Environment. 2017;**581**:782-793

[98] Saha KC. Chronic arsenical dermatoses from tube-well water in West Bengal during 1983-87. Indian Journal of Dermatology. 1995;**40**(1):1-12

[99] Bozack AK, Boileau P, Hubbard AE, Sillé FC, Ferreccio C, Steinmaus CM, et al. The impact of prenatal and early-life arsenic exposure on epigenetic age acceleration among adults in northern Chile. Environmental. Epigenetics. 2022;**8**(1):dvac014

[100] Chung YP, Chen YW, Weng TI, Yang RS, Liu SH. Arsenic induces human chondrocyte senescence and accelerates rat articular cartilage aging. Archives of Toxicology. 2020;**94**:89-101

[101] Caussy E, World Health Organization. A Field Guide for Detection, Management and Surveillance of Arsenicosis Cases. Regional Office for South-East Asia: World Health Organization; 2005. pp. 5-9

[102] Bhattacharya P, Welch AH, Stollenwerk KG, McLaughlin MJ, Bundschuh J, Panaullah G. Arsenic in the environment: Biology and chemistry. Science of the Total Environment. 2007;**379**(2-3):109-120

Section 3

Remedies

Chapter 7

Arsenic and Biosorption

Francisco Jose Alguacil and Jose Ignacio Robla

Abstract

Arsenic, either in (III) or (V) oxidation states forms, is a hazardous element to humans; thus, its removal from aqueous environments is of the utmost priority in the countries where this problem arises. From the various separation technologies, the removal of arsenic via biosorption processing attracted an interest, because besides the removal of the element, allows the recycle materials that in many cases are considered as wastes. The present chapter reviewed the most recent proposals (2022 year) about using biosorbents to remove this toxic element.

Keywords: biosorption, arsenic(III), arsenic(V), organoarsenic, water, wastewaters, desorption.

1. Introduction

Water contamination by heavy metals is one of the most serious risks to public health and the environment [1], being this contamination due to natural or anthropogenic causes. However, even these toxic elements can be present in flavored and functional drinking waters and in mineral waters [2].

Among these toxic metals, arsenic has a preeminent position due to its influence on living organisms, particularly on humans. The ingesta of this element is associated with a series of diseases, including cancer [3]; thus, its removal from waters or liquid effluents is of the utmost importance to prevent fatal consequences on the population. Moreover, this element is also associated to murder [4].

Arsenic can be present in aqueous solution in (III) or (V) oxidation states, and with both oxidation states, existed the possibility to find inorganic and organic As-bearing compounds; generally, organoarsenic is formed from inorganic arsenic by a process called biomethylation, and the toxicity of all these compounds depends of several issues. It is of general consent that the order to hazardousness of these arsenic compounds is ranked (from the more toxic to the less one) as: (i) inorganic As(III), (ii) organic As(III), (iii) inorganic As(V), and (iv) organic As(V). A comprehensive list of naturally occurring inorganic and organic As(III) and As(V) species is reported in the literature [5].

Conventional arsenic removal strategies, from metal-contaminated waters (including liquid effluents) included: oxidation, coagulation/precipitation, membrane filtration, adsorption, ion exchange, biosorption, and solvent extraction. Several recent literature reviewed the use of some of these technologies on this important toxicological issue [6–10]. The use of microbes in the remediation of arsenic from the environment is of particular interest, which is claimed as an eco-friendly and economical technology. To contend with arsenic, microorganisms have specific mechanisms such as biotransformation, biosorption, and homeostasis. Also, microbial approaches are mentioned in the remediation of metals from extraterrestrial materials, that is, meteorites [11]. Also, ferrate ($FeVIO_4^{2-}$) is now increasingly being synthesized and used as a novel adsorbent for the remediation of hazardous chemicals, including arsenic [12].

Among the above technologies, and due to the economic and relative easiness of operational modes, biosorption has a preeminent position in the removal of arsenic from contaminated waters. This chapter reviewed the literature published in 2022 about this important issue.

2. Arsenic(III)

Like in many metals involving systems, the removal of arsenic(III) from contaminated waters depends on its speciation in aqueous solutions. **Figure 1** shows this speciation at various pH values [13]. Obviously, the range of predominant species versus pH is a function of the arsenic(III) concentration in the solution, but it always follows the pattern as shown in the figure.

Accordingly to this distribution, arsenic(III) existed as soluble anionic species at alkaline pH values, whereas at acidic, neutral, or even slight alkaline pH values, $As(OH)_3$ or H_3AsO_3 appeared as predominant species; however, as it can be seen later, some investigations are performed at near neutral pH values, indicating the existence of As^{3+} cationic species at this pH range.

Cotton stalks-derived biochar (CSB) is an adsorbent material that has the potential to remove As(III) from aqueous solutions [14]. This material is characterized by different techniques: Brunauer–Emmett–Teller (BET), scanning electron microscopy (SEM), Fourier-transform infrared spectroscopy (FTIR), and point of zero charges (PZC) in order to known surface moieties that facilitate As(III) uptake onto the biochar. Using experimental conditions of 1 g/L biochar dosage, pH value of 6, reaction

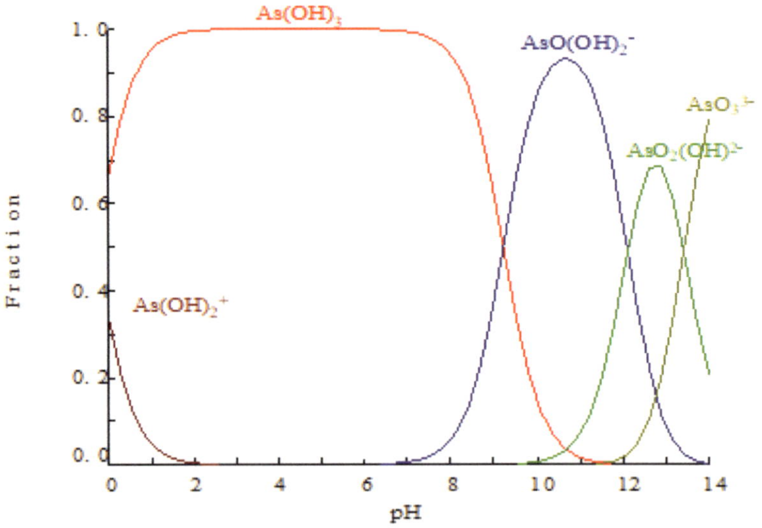

Figure 1.
Distribution of arsenic(III) species with the aqueous pH.

time of 2 h and an initial As(III) in the solution of 200 μg/L, experimental results revealed that the maximum As(III) adsorption capacity is 89.90 μg/g. The percentage of As(III) removal, from the solution, decreases with increasing As concentration in this phase. The linear form of the Langmuir model (Eq. (1), explains As(III) loading onto CSB, with maximum uptake of 102.78 μg/g (r^2 = 0.99). The linear form of the Langmuir isotherm is represented by the next equation:

$$\frac{[As]_{aq,e}}{[As]_{ba,e}} = \frac{1}{[As]_{ba,m} K_L} + \frac{1}{[As]_{ba,m}}[As]_{aq,e} \quad (1)$$

in the above equation, $[As]_{aq,e}$ and $[As]_{ba,e}$ are the arsenic concentration in the aqueous phase and in the adsorbent at the equilibrium, respectively, $[As]_{ba,m}$ is the maximum arsenic concentration in the bioadsorbent, and K_L is the Langmuir constant. As(III) uptake is attributed to high surface area of the bioadsorbent (103.62 m^2/g) and the presence of different functional groups, i.e. −OH, C=O, C−O on the CSB surface that facilitated As(III) adsorption and removal from the solution. Note from the authors of this review: in this work, AsO_3^{3-} is the responsible for metal uptake onto the bioadsorbent; however, at the pH value at which these experiments are done (pH 6), this species does not exist (see **Figure 1** of this review). Moreover, no desorption data are included in the work.

Next reference [15] isolated As-resistant lactic acid bacteria (LAB) and assessed the metal adsorption stoichiometry of LAB to validate its practical application as a bioremediation tool. In the investigation, 50 As-resistant LAB colonies from human (HS1–25) and albino mice (MM1–25) fecal samples are isolated. Considering these 50 As-resistant LAB, the HS12 isolate exhibited the highest arsenic removal efficiency (0.021 mg/h g). The metal bioremediation equilibrium investigation determined the contact time of 10 min and pH values in the 5–7 range for optimum arsenic biosorption from water. The non-linear Langmuir isotherm (r^2 = 0.993) fitted well with the experimental data. The arsenic bioaccumulation and scanning electron microscopy studies proved that, the mechanisms of metal sequestration of LAB HS12 is by binding of arsenic onto the cell membrane (0.000037 mg/g) as well as within the cell (0.000036 mg/g). The phylogenetic analysis of 16S rDNA amplicon (500 bp) of isolated potential HS12 LAB strains showed 97% similarity to *Lactobacillus reuteri*. Note from the authors of this review: no desorption data are included in the published manuscript.

A solid waste-based biosorbent derived from the *Cassia fistula* pod biomass is used to remove arsenic from aqueous solutions [16]. The prepared biosorbent has been characterized through different techniques, including field emission scanning electron microscopy, Fourier-transform infrared spectroscopy, and X-ray diffraction, to investigate the physiochemical properties of the material. The experiments have been performed considering four experimental variables: pH, biosorbent dosage, initial concentration of As^{3+}, and reaction time. The experimental results have been analyzed using the design-expert software for the optimization of different parameters. The maximum removal of arsenic (91%) could be achieved, whereas monolayer adsorption capacity is found to be 1.13 mg/g in 80 min at pH 6.0 and 30°C by using 60 mg dose of bioadsorbent. The arsenic adsorption behavior of the bioadsorbent has been well interpreted in terms of pseudo-first-order and Freundlich model. Note from the authors of this review: a major error occurs in this work since authors considered the existence of As^{3+} species in the aqueous phase, and this element (see **Figure 1**) does not form this cationic species at any aqueous pH value.

A metal oxide-based biosorbent for As(III) was designed and synthesized from eggshell biomass and characterized using FTIR, FE-SEM, EDX, XRD, and chemical analysis [17]. Raw eggshell (RES) powder is first dissolved with HCl and mixed with $ZrOCl_2{\cdot}8H_2O$ solution and then precipitated to obtain hydrated double oxide precipitate (HDOP), which is the material used for As(III) removal. Metal biosorption onto HDOP is fast and depends on pH and As(III) concentration. Pseudo-second order kinetics and Langmuir isotherm models (both in this linear form) are described As(III) uptake onto the bioadsorbent. HDOP provided exchangeable hydroxide/or chloride ligands for improved biosorption of As(III). The Langmuir model indicates that the maximum As(III) biosorption capacity onto HDOP is 40 mg/g at an optimum pH value of 10. Chloride and nitrate cause negligible interference on arsenic loading onto the adsorbent, whereas sulfate and phosphate significantly decreased the As(III) biosorption capacity of the biomaterial. HDOP completely removed arsenic from contaminated groundwater, and the remaining concentration reached values below the safe drinking water standard (10 μg/L) set by WHO. Metal desorption using three aqueous media, acid (1 M HCl), neutral (1 M NaCl), and alkaline (1 M NaOH), showed that in the alkaline medium, 94% removal rate from As-loaded bioadsorbent is reached, against 16% (NaCl) or 41% (HCl). After eight consecutive cycles of adsorption-desorption (with 2 M NaOH), results indicated that there is a continuous decrease in As(III) adsorption (99% in the first cycle) and 85% (fifth cycle), though the percentage of As(III) desorption always is greater than 95%.

A bioremediation study showed *Bixa orellana* as an accumulator of As(III) (and Cr(VI)); results are validated by SEM-EDX, FTIR, and other kinetic analyses [18]. Maximum percentage removal of As(III) is near 40% using an initial metal concentration of 6 mg/L. The bioadsorption data are well fitted to Freundlich and Elovich models. Three bacteria isolated from the coal mines of Rajmahal hills (India) showed As(III) resistance and bioremediation potentials (up to 150 mg/L). The 16S rRNA genotyping of these isolates is done (GenBank accession no: MK231250, MK23125, MK231251, and MK231252), which showed similarity with *Stenotrophomonas maltophilia*, uncultured gamma proteobacteria clone, and Bacterium E1. Further, the presence of genes involved in arsenic biotransformation like aox, acr, and ars is also confirmed in these bacterial isolates. Maximum percentage removal of As(III), from a 50 mg/L solution, by ASBBRJM16, ASBBRJM85, and ASBBRJM87 bacteria are 61%, 28%, and 10%, respectively. The isolated bacteria promoted the oxidation of As(III) to As(V) via the arsenic oxidase enzyme. Note from the authors of this review: desorption data are not presented in the manuscript.

Different biosorbents, from the weed *Rumex acetosella*, are used to remove metal cations in wastewater [19]. Drying, grinding, and sieving of the stems is carried out to obtain the biomass, retaining the fractions of 250–500 μm and 500–750 μm, which served to obtain the biosorbents in natura (unmodified), acidic, alkaline, and mixed. The 250 μm mixed treatment was the one that presented the highest removal percentages, mainly due to the OH, NH, −C−H, COOH, and C−O functional groups, achieving the removal of up to 33% of arsenic (also 96% for lead, 36% for zinc and 34% for cadmium). For contact times of 120 min and an optimum pH of 5.0, a loss of cellulose mass of 59% at 328°C and a change in the surface of the material are also observed, which allowed for obtaining topography with greater chelating capacity. Langmuir and pseudo-second-order models fitted the experimental data. Note from the authors of this review: this manuscript also presented the major error of considering the presence of As^{3+} in the solution. Moreover, no desorption data are included in the work.

Another type of exchanger is developed from waste biomass from watermelon rind after increasing the carboxyl functional groups by saponification [20]. Saponified watermelon rind (SWR) is further loaded with La(III) to attenuate the contamination of As(III) from water. As it is usual, different techniques are used to characterize the biosorbent. Arsenic speciation of adsorption product through X-ray photoelectron spectroscopic (XPS) analysis revealed that As(III) is partially oxidized to As(V) during the biosorption process. As(III) uptake onto La(III)-SWR is best described by Langmuir isotherm and pseudo-second-order kinetic model. Arsenic(III) removal from the solution is due to the exchange of $H_2AsO_3^-$ from the solution and OH^- groups of the bioadsorbent:

$$P \equiv La - OH + H_2AsO_3^- \rightarrow P \equiv La - H_2AsO_3 + OH^- \quad (2)$$

where P represented the non-reactive part of the bioadsorbent. At a pH of 12.08, the optimum biosorption capacity is found to be 38, 49, and 63 mg/g, at temperatures 25°C, 30°C, and 35°C, respectively. The presence of chloride and nitrate in the solution has negligible interference with As(III) adsorption, whereas sulfate and phosphate significantly decrease As(III) biosorption. A thermodynamic study showed the spontaneous and endothermic nature of As(III) biosorption onto La(III)-SWR. As it is noted above, As(III) partially oxidized to As(V) due to aerial oxidation occurring during the adsorption process; in any case, $H_2AsO_4^-$ species also is bioadsorbed onto La(III)-SWR bioadsorbent. Elution is carried out with NaOH solutions, increasing the percentage of arsenic removal from the bioadsorbent when the NaOH concentration increases from 0.01 M to 2 M (near 98% arsenic removal).

To investigate arsenic biosorption by *Chondracanthus chamissoi* and *Cladophora* sp. macroalgaes [21], approximately 5 kg of algae are collected from Huanchaco's beach and Sausacocha lake (Huamachuco, Peru), La Libertad, and used in the study [21]. Arsenic biosorption is carried out in four column systems, with 2 g of algae in pellets form each, circulating arsenic solutions of 0.25 and 1.25 mg/L at 300 mL/min cm^2. Metals concentration is determined at 3 and 6 h of treatment. At 6 h, *Chondracanthus chamissoi* presented an As biosorption of 96% and 85% from the 0.25 mg/L and 1.25 mg/L solutions, respectively. At the same time, *Cladophora* sp. presented arsenic biosorption of 96% and 42% from the above solutions. It is concluded that *Chondracanthus chamissoi* achieves higher percentages of biosorption than *Cladophora* sp. from solutions of 1.25 mg/L As, and that there is no significant difference between the biosorption percentages of Chondracanthus chamissoi and Cladophora sp. from 0.25 mg/L solutions of the metal after 6 h of treatment. In the investigation, the algae are pretreated with HCl solutions, since this pretreatment improves the biadsorption capacity of the algae. This improvement is due to the protonated form, and there is a release of protons that favors the exchange of metals from the solution. Note from the authors of this review: this release of protons should be indicative of a cation exchange reaction; thus, do the authors consider once more that arsenic(III) is present as a cation in the solution? No desorption data are included in the work.

Magnetite nanoparticles (MNPs) are synthesized using the seaweed–*Ulva prolifera*, an amply found marine source in the Western coastal regions of India [22]. This biosorbent material is used as nanoadsorbent for the removal of As(III) from an aqueous solution. The optimum conditions to achieve the best arsenic(III) removal (97.5%) from the solution are: pH 9, rotation speed of 150 rpm, 90 min of reaction time, and 1.15 g/L MNPs dosage. The uptake process fitted with the non-linear forms of Langmuir isotherm and pseudo-second-order kinetic model. The highest arsenic adsorption capacity is 12 mg/g in a spontaneous and endothermic process. Note from the authors of this review: desorption experiments are absent in work.

Feather keratin-derived biosorbents using water-dispersed graphene oxide are used to create a biadsorbent for arsenic(III) [23]. Cross-linking of feather keratin with graphene oxide is investigated through X-ray photoelectrons spectroscopy (XPS), scanning and transmission electron microscopy, and Brunauer-Emmett-Teller (BET) analysis. The modifications resulted in increased surface area of the keratin proteins with substantial morphological changes, including the development of cracked and rough patches on the surface. These biosorbents exhibited excellent characteristics for the simultaneous removal (up to 99% in 24 h) of metal oxyanions, including arsenic, selenium, chromium, and cations including nickel, cobalt, lead, cadmium, and zinc from polluted synthetic water containing 600 μg/L of each metal. The insights into the biosorption mechanism revealed that the electrostatic interaction, chelation, and complexation primarily contributed to the removal of multiple heavy metal ions in a single treatment. Note from the authors of this review: desorption data are not included in the work.

Dry microalga *Chlamydomonas* sp. is used to remove arsenic(III) from aqueous solutions [24]. The experimentation was performed at a constant temperature of 25°C and a shaking speed of 300 rpm. Best conditions for As(III) removal are: pH 4, contact time of 60 min, temperature 25°C, and biomass concentration of 0.6 g/L. Metal uptake follows the Langmuir isotherm in its linear form, being the adsorption process of the endothermic character. Several desorbents are used (0.1 N solutions of HNO_3, H_2SO_4, HCl, NaOH, and EDTA), with the best desorption percentages yielded when EDTA is used, being this attributable to the chelating properties of this acid. Again, there is a continuous decrease in adsorption capacity after continuous use, that is, 76% removal capacity in the first cycle against 65% in the fifth cycle.

A recyclable biofilm material is prepared by loading *Herminiimonas arsenicoxydans* (*H. arsenicoxydans*) onto electrospun biomass-activated carbon nanofibers [25]. The removal of arsenic(III) is accompanied by catalytic oxidation of this oxidation state to As(V), being this oxidation attributable to a large amount of biomass accumulated and the formation of biofilms on the surface of the biomass-activated carbon nanofibers. The oxidation process responded to the first-order kinetic model. After five consecutive runs, the biofilms presented a convenient recyclability.

An adsorbent based on hydrous ferric oxide-impregnated agarose beads is used to remove As(III) from pharmaceutical wastewater [26]. The adsorption process fits well with the Langmuir model with nearly 76 mg/g of maximum arsenic load. The system is implemented in a column packed with the adsorbent beads, decreasing an initial arsenic concentration of 250–10 μg/L after five continuous cycles. Desorption is accomplished with NaOH solutions in the 0.05–3 M range, though column experiments used 0.1 M NaOH solutions as a compromise between desorption kinetics and rate of arsenic desorption. Continuous use does not affect the beads morphology. Whereas arsenic(III) is loaded onto the beads via and inner and outer sphere complexation-association, desorption is mainly governed by the removal of the outer sphere arsenic species.

Rice husk is the precursor of an eco-friendly Fe-Al bimetallic oxide/biochar adsorbent composite used in the removal of arsenic(III) from aqueous medium [27].

Maximum adsorption is reached at pH 5.5, being the experimental adapt fitted to the Freundlich isotherm and the pseudo-first-order kinetic model. Desorption is accomplished with 0.5 M NaOH solutions; however, there is a continuous decrease in the adsorption rate after continuous use, that is, 75% in the first cycle and 30% after the fifth cycle; this decrease is attributable to (i) incomplete desorption after each cycle and (ii) loss of adsorbent material in the recycling process.

Psidium guajava (guava) leaf has been investigated as effective adsorbent for As(III) [28]. Different experimental variables are considered, and results show that regardless of arsenic concentration (0.01–0.05 g/L), the rate of arsenic(III) removal from the solution reached a maximum at pH 10, and decreased from pH 10 to pH 2. Also, it is shown that the percentage of arsenic(III) removal from the solution decreased as the initial metal concentration in the solution increased from 0.01 to 0.05 g/L. Several solutions, at 0.1 N concentration, are tested to desorb arsenic loaded onto the adsorbent, and the HCl medium presented the best rate of desorption if compared with sulfuric and nitric acid, sodium hydroxide, and water.

A bionanocomposite bead UiO-66/CB was fabricated through simple gelation followed by the freeze-casting method and used to decrease the concentration of As(III) from contaminated water [29]. The biocomposite was fabricated by dispersion of UiO-66 nanoparticles in networking pores of macroporous chitosan beads. Moreover, the millimeter-sized UiO-66/CB-2.0 performed better, with respect to arsenic(III) removal, than the freely dispersed UiO-66 nanoparticles. 1 g/L UiO-66/CB-2.0 reduced As(III) concentrations from 200 μg/L in groundwater to below 10 μg/L. Arsenic(III) can be desorbed with 0.1 M NaOH solutions, though a slight reduction in the adsorption efficiency was observed after five cycles. The investigation also presented data on the use of the modified biochar in a fixed-bed column.

Dithiocarbamate (DTC)-modified cellulose adsorbents can selectively separate metal ions from water, though they can be dissolved in this medium. Trying to resolve this problem, the adsorbent was cross-linked with epoxy or complexed with iron [30]. The iron-complexed adsorbents still had solubility problems, but cross-linkage with 6.0 mol% of epoxy resulted in a material that was almost insoluble and dispersed in the solution. The optimum contact time and pH for As(III) removal were 20 min and 3.0, respectively; the adsorption responded to the Langmuir isotherm equation, being the adsorption attributed to the next reaction:

$$3DTC + As(OH)_3 = As(DTC)_3 + 3OH^- \qquad (3)$$

thus, the adsorption process was increased at acidic pH values. Note from the authors of the review: apparently, the authors of the manuscript did not interpret well the adsorption mechanism, Eq. (3), because it was doubtful that $As(OH)_3$ or H_3AsO_3 species predominant at these pH values (see **Figure 1**) lost OH^- groups; moreover, at these acidic pH values OH^- species never exists. No desorption data are included in the manuscript.

Using biochar derived from cotton stalks, the next reference [31] used a chemical treatment to modify the adsorption properties of the biomaterial. Thus, in the investigation, three biochars were used in the removal of arsenic from contaminated As(III) skipped-drinking water: CSB (cotton stalk biochar), HN-CSB (treated with nitric acid, and Na-CSB (treated with sodium hydroxide). An isotherm model showed that arsenic biosorption was best fitted to the Langmuir isotherm in all the three biochars

CSB (qmax = 103 μg/g), Na-CSB (qmax = 151 μg/g), and HN-CSB (qmax = 157 μg/g). The chemical modification of the biochar produced biomaterials with the largest surface area, porous structure than the pristine material, and also promoted the presence of new functional groups on the surface of the modified biochars. Note from the authors of the review: desorption data were not given in the manuscript.

3. Arsenic(V)

Similarly to arsenic(III), arsenic(V) presented different speciation at the various pH values of the aqueous solution. This speciation is shown in **Figure 2** [32], which demonstrates the sequence of neutral to anionic arsenic(V) species as the pH of the solution increased. Thus, the bioadsorption of this element presumably will be dependent, among other variables, on this speciation, and thus, of the pH of the aqueous phase.

The removal of As^{5+} in cassava wastewater, using an efficient biosorbent from chemically pretreated unshelled *Moringa oleifera* seeds, is investigated [33]. Several experimental conditions are used to investigate, in order to achieve efficient metal removal from the wastewater. The results of Fourier-transform infrared spectroscopy clearly suggested that additional functional groups attributed to esters are formed in the pretreated biosorbent, which is responsible for the improvement in biosorption. It was found that contact time, bioadsorbent dosage, and bioadsorbent pretreatment concentration had a significant influence on arsenic removal. Maximum percentage removal of 99.9% is reached in the synthetic solution at pH 4.0, contact time of 30 min, and dosage of 2 g for biosorbent pretreated with 1 M of chemical solution. It is shown that the bioadsorption process for untreated bioadsorbent is by ion exchange, while for treated material, the removal of the metal is attributable to a multifarious adsorption mechanism. The biosorption process was exothermic and spontaneous. Note from the authors of this review: once again, a major error occurs

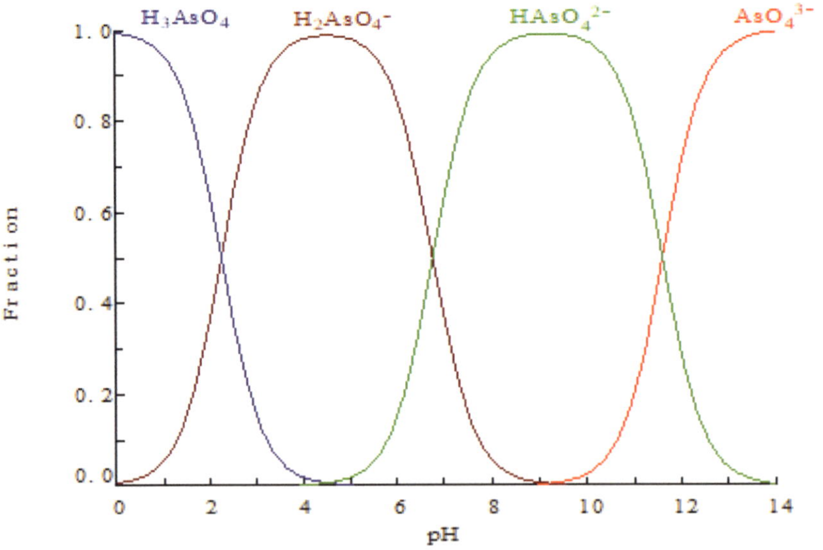

Figure 2.
Distribution of arsenic(V) species with the aqueous pH.

since the authors considered the existence of cationic As^{5+} species in the aqueous solution (also chromium(VI) as Cr^{6+}), that evidently (see **Figure 2**) does not exist. No desorption data are included in the manuscript.

The treatment of arsenic-contaminated water with low-cost and environmental-friendly adsorbents, such as biochar, is considered a promising technique [34]. Thus, the treatment of As-contaminated water using eggshell biochar was investigated. Various parameters affecting the adsorption process, such as pH, contact time, adsorbent dosage, As(V) concentration, and the effects of anions, are studied. The results revealed that at a pH of 4.5, maximum adsorption of near 6 mg/g is reached (after 2 h), with a percentage of As(V) removal of 96%, from an initial metal concentration of 0.6 mg/L and a bioadsorbent dosage of 0.9 g/L. The SEM-EDS data illustrated that biochar consisted of a large number of active sites for As(V) adsorption, appearing the metal species on the biochar surface after metal uptake. This uptake is well represented by the Freundlich and the pseudo-second-order kinetic models. The presence of phosphates in the solution is detrimental for arsenic(V) loading onto the biochar, this being probably attributable to the fact that phosphate ions are adsorbed, on the same adsorption sites of the biochar, as arsenic(V) species. Note from the authors of this review: The manuscript does not include desorption data.

Artocarpus heterophyllus seed powder is used as a biosorbent material to remove different heavy metals [35]. The batch adsorption studies confirmed the higher removal percentage of this bioadsorbent (jackfruit) seed powder for arsenic (As^{5+}), cadmium (Cd^{2+}), and chromium (Cr^{6+}), while lower efficiency is observed for other heavy metals like copper (Cu^{2+}), zinc (Zn^{2+}), and nickel (Ni^{2+}). Optimization of various process parameters is carried out, and the optimum conditions are: adsorbent weight of 0.5 g for initial metal concentrations of 40 μg/L, 30 mg/L, and 30 mg/L; contact time of 10 h, 8 h, and 6 h; process temperature from 25 to 30°C; pH of 7, 7.5, and 7.5 for arsenic, cadmium, and chromium, respectively. The equilibrium data of the study are well fitted for Langmuir isotherm (arsenic, cadmium, and chromium) and Freundlich isotherm (arsenic and chromium). The kinetic and thermodynamic study confirmed that the adsorption of all three heavy metals followed the pseudo-second-order kinetics in an endothermic and spontaneous process. Several solutions are considered to desorb arsenic(V) from an As-loaded bioadsorbent. The best results are obtained when 1 M HCl solutions are used. However, after various adsorption–desorption cycles, a continuous decrease of the adsorption is observed: near 70% in the first cycle and 20% in the fifth cycle. Note from the authors of this review: again, authors considered arsenic(V) species as the non-existent As^{5+}, and the same major error occurs in the case of chromium(VI) speciation: consideration of the existence of Cr^{6+} cation in the aqueous solution.

An adsorbent based in a matrix of diatomitex and chitosan as the modifier is used to investigate its properties to remove As(V) from solutions [36]. In the hybrid adsorbent, open chains of chitosan are grafted onto the matrix surface, and arsenic is loaded onto the adsorbent through −NH2 groups of the biomaterial. Maximum arsenic(V) removal from the solution occurs at pH 5, presenting the hybrid adsorbent with better performance, with respect to As(V) removal, than the pristine diatomite. Desorption is investigated using HCl solutions, with a slight decrease in As(V) uptake onto the adsorbent after continuous cycles, 94.2% in the first cycle and 90.4% in the fifth cycle.

Chitosan, supported onto a modified polypropylene membrane, is another bioadsorbent used to remove arsenic(V) from aqueous solutions [37]. The removal of arsenic from the solutions depended on the pH of the solution and the degree in which chitosan is grafted to the membrane. Metal adsorption, at pH 6.5, is related to a chemisorption process:

$$R-NH_3^+ + H_3AsO_4 \rightarrow R-NH_3^+H_2AsO_4^- + H^+ \quad (4)$$

obeying the pseudo-second-order kinetic model equation. Note from the authors of this review: accordingly to the diagram of distribution (**Figure 2**), at this pH of 6.5, H_3AsO_4 is not the predominant As(V) species in solution. No desorption data is included in the work.

Miscanthus biochar has been used to investigate its performance on As(V) (and Cd(II) from aqueous media of different pH values [38]. Arsenic(V) is best removed from the solution at alkaline pH values; in the removal process, arsenic(V) is reduced to arsenic(III) and zero-valent arsenic. This reduction being attributable to electrons from the biochar under a physisorption and hydrophobic interactions between arsenic and the adsorbent. Note from the authors of this review: as it is noted in the manuscript, the process leads to an adsorbent loaded with arsenic(III), which represented a further hazard to humans and environment. No desorption data included in the work.

A composite of MgO nanoparticles with Itsit biochar (MgO-IBC) was used to remove arsenate from contaminated water [39]. Experimental results showed that both pH of the water and temperature influenced the overall adsorption efficiency of the biocomposite: low pH and high temperature resulted in higher arsenate bioadsorption. The removal process fitted the pseudo-second-order model and the Langmuir isotherm, through a chemisorption and spontaneous process. The arsenic-loaded adsorbent was washed with 6% HCl solution; however, the percentage of arsenic adsorption dramatically decreased from the first (96%) to the fourth cycle (32%).

Two biocomposites: iron(III)-chitosan and iron(III)-chitosan-CTAB composites, were prepared using an ionotropic gelation method [40]. Compared with the former composite, iron(III)-chitosan-CTAB material was more effective for As(V) adsorption at a wide range of pH (3.0–8.0). Again, adsorption processes fitted well with the pseudo-second-order model, though in these cases, the removal of arsenic responded to the Freundlich isotherm. In the presence of $H_2PO_4^-$, the adsorption of As(V) decreased. 1% w/v NaOH solution was the best desorption agent. Both bioadsorbents maintained their initial adsorption capacities after five adsorption-desorption cycles. Characterization results indicated that both electrostatic attraction and surface complexation mechanisms were responsible for arsenic(V) uptake ($H_2AsO_4^-$) onto these two bioadsorbents.

4. Biosorption systems for As(III) and As(V)

A magnetic biochar is generated by pyrolysis of waste leaves of *Raphanus sativus* or *Artocarpus heterophyllus* peel pretreated with FeCl3 and used to remove As(III) and As(V) from solutions [41]. Maximum uptake onto the two adsorbent is very similar for both oxidations states, that is, 2.08 mg/g and 2.03 mg/g for As(III) and As(V), respectively. The pseudo-second-order kinetic model fitted well with the adsorption of both arsenic oxidation states and the two bioadsorbents. Note from the authors of this review: no desorption data included in the manuscript.

Calcium alginate caged graphene oxide functionalized with metformin (Alg@GMet) forming beads is used to investigate the removal from aqueous solutions, both for As(III) and (V) [42]. Best removal rate depended on the arsenic oxidations state: (i) in the (III) state maximum loading resulted in pH values of 2–6, being H_3AsO_3 the loaded arsenic species, (ii) arsenic(V) is loaded at pH values in the 2–4 range, being H2AsO4- the species removed from the solution. In both cases, the adsorption process fitted well with the linear form of the Langmuir isotherm and the pseudo-second-order kinetic equation. Also, in both cases, the adsorption process is exothermic, indicating the values of $\Delta G°$ that in both cases, the adsorption responded to a physisorption process. Desorption is accomplished with 0.1 M NaOH; however, there is a loss of adsorption capacity, that is, As(III) from 81 to 64%, As(V) from 98 to 62%, being the values from the first to the seventh cycle. The system is used in the removal of arsenic from tap water and sewerage waste; removal rates are: As(III) 69% (tap water) and 73% (sewerage waste), As(V) 62% (tap water), and 66% (sewerage waste).

Two activated biochar materials: peanut char (δ-MnO_2/A-PC) and corn char (δ-MnO_2/A-CC), were used in the treatment of a solution containing 97.5% As(III) and 2.5% As(V) [43]. Using δ-MnO_2/A-PC, 18.8% of As(III) and 35.4% of As(V) remained in the solution after 24 h of contact between both phases, indicating that (i) part of As(III) was removed from the solution by the adsorption process and (ii) that As(III) was also partially oxidized to As(V). Experimental results indicated that δ-MnO_2/A-CC was more suitable for removing waters with low As(V) concentrations, whereas δ-MnO_2/A-PC performed betters in solutions containing high As(III) concentrations. Fourier-transform infrared spectroscopy and X-ray diffraction analyses demonstrated that δ-MnO_2 was coated onto the surfaces of the biochars. Note from the authors of the review: no desorption data are included in the work.

The next investigation [44] utilized Fe^{2+} ($FeCl_2$) and zirconium oxychloride ($ZrOCl_2$) to synthesize a modified biochar (FeZrO-BC) by a co-precipitation method. The biochar was used to remove arsenite and arsenate from contaminated natural water. The addition of these two metals resulted in the formation of positively charged Zr-O and Fe-O groups. Kinetics and isothermal adsorption results indicated that both arsenic species are removed from the water via chemisorption on a monolayer, with maximum adsorption capacities of 46.7 mg/g As(III) and 47.8 mg/g As. Water was ineffective in desorbing arsenic-loaded materials, though both As(III) and As(V) can be desorbed with 0.05 M NaOH or nitric acid media. There was a loss in adsorption capacity from the first to the third cycle (**Table 1**).

Iron(III) chloride impregnation of bagasse fly ash produced a biomaterial used in the removal of As(III) and As(V) from aqueous solutions [45]. Maximum removal of As(III) (95%) and As(V) (97%) was yielded from solutions containing less than

Cycle	NaOH	HNO_3
First	89.7	89.4
Second		
Third	49.3	55.4

Data for As(III). As(V) adsorption performed the same.

Table 1.
FeZrO-BC loss of capacity after continuous use.

20 μg/dm3), whereas with solutions of 500 μg/dm3, the percentage of removal was 86% (As(III)) and 87% (As(V)) 3 g/dm3. The adsorption of both arsenic species responded to the pseudo-second-order kinetic model in an exothermic and spontaneous process. The regeneration study was carried out by different solvent and thermal methods; with HCl, the best desorption results were yielded: As(III) (83%), As(V) (74%). Thermal desorption produced a continuous decrease in the adsorption capacity, obviously due to a degradation of the adsorbent.

A magnetite-impregnated nitrogen-doped hybrid biochar (N/Fe_3O_4@BC) was used for the removal of arsenate and arsenite from aqueous environment [46]. Maximum monolayer adsorption values were 18.15 mg/g (arsenate) and 9.87 mg/g (arsenite), which were higher values than that of pristine biochar 9.89 and 8.12 mg/g, respectively. Adsorption fitted to the pseudo-first-order model, indicative of a physicochemical process. The bioadsorption of arsenic species was attributable to the presence of surface groups (OH−, $-NH_2$, and −COOH), electrostatic attraction (via H- bonds), surface complexation and ion exchange followed by external mass transfer diffusion, and As(III) oxidation into As(V) by (N/Fe_3O_4@BC) reactive oxygen species. NaOH medium was used to desorb As(III) and As(V) from the loaded adsorbent, and it was observed a continuous loss in adsorption properties from the first (90% As(III), 80% As(V)) to the seventh cycle (60% As(III), 55 As(V)). Also, some degree of arsenic oxidation was observed in the desorption step.

5. Biosorption systems without mentioning the metal oxidation state

The next reference [47] investigates the combined photosynthetic activities of two green microalgal species, *Tetradesmus obliquus* and *Tetradesmus reginae*, on an integrated biophotovoltaic (BPV) platform for simultaneous wastewater treatment, toxic metal biosorption, carbon biofixation, bioelectricity generation, and biodiesel production. The wastewater used in the experimentation is collected at the Daspoort Wastewater Treatment Plant (DWWTP) located at the Tshwane Metropolitan Municipality in Central Pretoria, South Africa. The experimental setup comprised a dual-chambered BPV with copper anode surrounded by *T. obliquus* in BG11 media, and copper cathode with *T. reginae* in municipal wastewater separated by Nafion 117 membrane. The investigation reported a maximum power density of 0.344 W/m^2 at a cell potential of 0.415 V with external resistance of 1000 Ω and 0.3268 V maximum open-circuit voltage. The wastewater electrical conductivity and pH increased from 583 ± 22 to 2035 ± 29.31 mS/cm and 7.4–8.3, respectively, signifying increased photosynthetic and electrochemical activities. Residual nitrogen, phosphorus, chemical oxygen demand, arsenic, cadmium, chromium, and lead removal efficiencies by *T. reginae* are 100%, 81%, 72%, 48%, 89%, 71%, and 93%, respectively. Note from the authors of this review: no adsorption data were included in the work.

Natural waste adsorbents are used to eliminate arsenic and cadmium from aqueous solutions, and at the same time, reducing the amount of waste products [48]. The adsorbents used in the investigation are coconut husk and banana peel. Different experimental conditions are used in the work: adsorbent dosages (0.1–0.3 g), contact time (30–70 min), and temperature (25–45°C). The FTIR analysis revealed that certain heavy metals are more likely to load onto these adsorbents due to the presence of −OH and C=O functional groups. The optimum removal conditions are 0.1 g dose of adsorbent, and 70 minutes of contact time at a temperature of 25°C. The results revealed that banana peel removed 0.148 mg/L of arsenic (0.948 mg/L of cadmium) from the aqueous solution, suggesting that it is a more efficient adsorbent than coconut husk. Using banana peel as bioadsorbent, the percentage of arsenic removal is in the 8–22% range (94–99% for cadmium). Note from the authors of this review: no adsorption data were included in the work.

Apple residues, banana peel, eggshell, potato peel, and sweet potato peel were tested as biosorbents to remove toxic metals (As, Cd, Hg, and Pb) from contaminated waters [49]. Adsorption experiments are performed using 0.5 g of each biosorbent in 1 L of natural tap water spiked with a mixture of the above metals at realistic concentrations (50 μg/L) under different pH values (4.5, 6.5, and 8.5) and water salinities (0, 10, and 30). The analysis by scanning electron microscopy showed differences among biosorbents, mainly in pore size and fibrous structures. Fourier-transform infrared spectroscopy identified cellulose, hemicellulose, pectin, and lignin in all biosorbents, except in eggshells, which are constituted mainly by carbonates. Results showed that the levels of all the above metals in water are considerably reduced by the biosorbents and in less than 3 h (more accentuated for apple and banana peel). However, with all the above bioadsorbent the percentage of arsenic removal never is greater than 6%, value much lower than those of mercury (up to 99%), cadmium (76%), and lead (86%). Note from the authors of this review: again, desorption data are not included in the manuscript.

6. Organoarsenics

This review manuscript [50] considered a series of organoarsenics (roxarsone, p-arsalinic acid, carbazone, triphenylarsine, phenylarsenic acid, etc.) to describe some features in their removal from aqueous solutions. Data collected included the effect of the pH, temperature, initial concentration of the organoarsenics, adsorbent dosages, contact time, and other variables on organoarsenics removal from solutions. The manuscript also insights into the mechanism of organoarsenics uptake onto various adsorbents and also the fit of experimental data on loading isotherms and kinetics models (these adsorptions generally fit better to the Langmuir isotherm and the pseudo-second-order kinetics model). Finally, some data about desorption and regeneration operations are included in the manuscript. Generally speaking, there is a progressive decline of active sites during continuous adsorption-desorption cycles that could be responsible of this decrease in adsorbent performance.

Biochar composites fabricated from polyaluminum chloride (PAC) sludge were used to investigate their adsorption properties toward dimethylarsinic acid (DMA) and also inorganic As(III) and As(V) [51]. In the case of DMA, the removal process fitted the pseudo-second-order kinetics and Freundlich isotherm models. In the adsorption process, DMA suffered a demethylation process, and As(V) was reduced to As(III) due to microorganisms present in the system. A large-scale field experiment carried out in an artificial ecological wetland showed that the addition of biochar reduced the total arsenic concentration to be immobilized in wetland sediment by 19%. *Note from the authors of the review: desorption data were not included in the manuscript.*

7. Conclusions

This work reviews the most recent additions to the study of arsenic removal from aqueous solutions using bioadsorbents. This technology appears to be the most used separation technology, to investigate the removal of this hazardous metal, in comparison with other recently published data about the use of other technologies in this important issue.

It is worth noting that these investigations mainly focus on scientists in Africa and Asia. An analysis of the works reviewed here, from ref. [14–31] and [33–51], 66.7% of the manuscripts are generated from Asia (India, Nepal, Pakistan, Saudi Arabia, Malaysia, China, Iran, Republic of Korea, and Taiwan), 18.5% from Africa (Egypt, Nigeria, and South Africa), 7.4% from South America (Peru) and 3.7% each from North America (Canada) and Europe (Portugal) (the classification is done in the basis of the institution in which the corresponding author is located).

Despite the great interest raised within these investigations, these reviewers found several important drawbacks in some of the published manuscripts:

in a number of published manuscripts, authors consider that both arsenic(V) and arsenic(III) elements are present, in the aqueous solutions, as the respective cationic species As^{5+} and As^{3+}, which is a major chemical error since these cations never exist in aqueous solutions The above decreases the potential interest of the respective published works,

a number of the published manuscripts, 17 from a total of 27, which means 63%, do not consider the desorption step. This can also be considered an error, since the removal of a given element via biodesorption (or other separation technology) always consists of two steps: metal uptake onto the bioadsorbent-desorption, thus, without knowledge of how the potential biadsorbent performs in the desorption step, one has not an overall view of the whole removal process,

in the case in which this desorption step is considered, not a single manuscript mentioned what to do with the desorbed solution, which presumably contains a greater arsenic concentration than the original feed solution, resulting in a more toxic waste.

The potential of bioadsorbents, to remove arsenic from solutions, is here and is of current interest (both as a research tool and in a practical form), and hopefully, some of the drawbacks mentioned above will be resolved in future publications.

Besides these potential approaches to remove this contaminant from waters, governments must still apply strict laws to remediate this issue; if not, life together to the contamination problem go on.

Acknowledgements

The authors thank CSIC (Spain) for Project 202250E019.

Conflict of interest

The authors declare no conflict of interest.

DOI: http://dx.doi.org/10.5772/intechopen.1001315

Author details

Francisco Jose Alguacil* and Jose Ignacio Robla
Centro Nacional de Investigacions Metalurgicas (CSIC), Madrid, Spain

*Address all correspondence to: fjalgua@cenim.csic.es

References

[1] Zakeri HR, Yousefi M, Mohammadi AA, Baziar M, Mojiri SA, Salehnia S, et al. Chemical coagulation-electro Fenton as a superior combination process for treatment of dairy wastewater: Performance and modelling. International Journal of Environmental Science and Technology. 2021;**18**:3929-3942. DOI: 10.1007/s13762-021-03149-w

[2] Lorenc W, Markiewicz B, Kruszka D, Kachlicki P, Barałkiewicz D. Total versus inorganic and organic species of As, Cr, and Sb in flavored and functional drinking waters: Analysis and risk assessment. Molecules. 2020;**25**:1099. DOI: 10.3390/molecules25051099

[3] Monteiro De Oliveira EC, Siqueira Caixeta E, Santana Vieira Santos V, Barbosa Pereira B. Arsenic exposure from groundwater: Environmental contamination, human health effects, and sustainable solutions. Journal of Toxicology and Environmental Health Part B. 2021;**24**:119-135. DOI: 10.1080/10937404.2021.1898504

[4] Emsley J. The Elements of Murder. Oxford: Oxford University Press; 2005

[5] International Agency for Research on Cancer. Arsenic, metals, fibres and dusts. IARC Monograph on the evaluation of carcinogenic risks to humans [Internet]. 2012. Available from: https://www.ncbi.nhm.nih.gov/books/NBK304380 [Accessed: January 13, 2023]

[6] Ayub A, Srithilat K, Fatima I, Panduro-Tenazoa NM, Ahmed I, Akhtar MU, et al. Arsenic in drinking water: Overview of removal strategies and role of chitosan biosorbent for its remediation. Environmental Science and Pollution Research. 2022;**29**:64312-64344. DOI: 10.1007/s11356-022-21988-z

[7] Rai PK. Novel adsorbents in remediation of hazardous environmental pollutants: Progress, selectivity, and sustainability prospects. Cleaner Materials. 2022;**3**:100054. DOI: 10.1016/j.clema.2022.100054

[8] Singh A, Chauhan S, Varjani S, Pandey A, Bhargava PC. Integrated approaches to mitigate threats from emerging potentially toxic elements: A way forward for sustainable environmental management. Environmental Research. 2022;**209**:112844. DOI: 10.1016/j.envres.2022.112844

[9] Somayaji A, Sarkar S, Balasubramaniam S, Raval R. Synthetic biology techniques to tackle heavy metal pollution and poisoning. Synthetic and Systems Biotechnology. 2022;**7**:841-846. DOI: 10.1016/j.synbio.2022.04.007

[10] Srivastava V, Karim AV, Babu DS, Nidheesh PV, Kumar MS, Gao B. Metal-loaded biochar for the removal of arsenic from water: A critical review on overall effectiveness, governing mechanisms, and influential factors. Chemistry Select. 2022;**7**:e202200504. DOI: 10.1002/slct.202200504

[11] Choudhury S, Chatterjee A. Microbial application in remediation of heavy metals: An overview. Archives of Microbiology. 2022;**204**:268. DOI: 10.1007/s00203-022-02874-1

[12] Wei L, Lu J. Adsorption of microcystin-LR by rice straw biochars with different pyrolysis temperatures. Environmental Technology and Innovation. 2021;**23**:101609. DOI: 10.1016/j.eti.2021.101609

[13] Alguacil FJ, Escudero E. The removal of toxic metals from liquid effluents

by ion exchange resins. Part VIII: Arsenic(III)/OH-/Dowex 1x8. Revista de Metalurgia. 2018;**54**:e132. DOI: 10.3989/revmetalm.132

[14] Ahmad I, Farwa U, Khan ZU, Imran M, Khalid MS, Zhu B, et al. Biosorption and health risk assessment of arsenic contaminated water through cotton stalk biochar. Surfaces and Interfaces. 2022;**29**:101806. DOI: 10.1016/j.surfin.2022.101806

[15] Bhakta JN, Ohnishi K, Tsunemitsu Y, Ueno D, Manna I. Assessment of arsenic sorption properties of lactic acid bacteria isolated from fecal samples for application as bioremediation tool. Applied Water Science. 2022;**12**:116. DOI: 10.1007/s13201-022-01634-2

[16] Giri DD, Jha JM, Srivastava N, Hashem A, Abd A, Elsayed F, et al. Sustainable removal of arsenic from simulated wastewater using solid waste seed pods biosorbents of *Cassia fistula* L. Chemosphere. 2022;**287**:132308. DOI: 10.1016/j.chemosphere.2021.132308

[17] Gyawali D, Bhandari S, Basnet P, Dahal B, Upadhyaya IR, Ghimire KN, et al. Synthesis and characterization of metal oxide based ion exchanger from chicken egg shell biomass for the removal of arsenic from water. Sustainable Chemistry and Pharmacy. 2022;**30**:100870. DOI: 10.1016/j.scp.2022.100870

[18] Kumar U, Singh RS, Mandal J, Nayak AK, Jha AK. Removal of As(III) and Cr(VI) from aqueous solutions by Bixa orellana leaf biosorbent and As(III) removal using bacterial isolates from heavy metal contaminated site. Journal of the Indian Chemical Society. 2022;**99**:100334. DOI: 10.1016/j.jics.2021.100334

[19] Ligarda-Samanez CA, Choque-Quispe D, Palomino-Rincón H, Ramos-Pacheco S, Moscoso-Moscoso E, Huamán-Carrión ML, et al. Modified polymeric biosorbents from *Rumex acetosella* for the removal of heavy metals in wastewater. Polymers. 2022;**14**:2191. DOI: 10.3390/polym14112191

[20] Lochan Aryal R, Thapa A, Raj Poudel B, Raj Pokhrel M, Dahal B, Paudyal H, et al. Effective biosorption of arsenic from water using La(III) loaded carboxyl functionalized watermelon rind. Arabian Journal of Chemistry. 2022;**15**:103674. DOI: 10.1016/j.arabjc.2021.103674

[21] Otiniano NM, De La Cruz-Noriega M, Cabanillas-Chirinos L, Rojas-Flores S, Muñoz-Ríos MA, Rojas-Villacorta W, et al. Arsenic biosorption by the macroalgae Chondracanthus chamissoi and Cladophora sp. PRO. 2022;**10**:1967. DOI: 10.3390/pr10101967

[22] Selvaraj R, Murugesan G, Rangasamy G, Bhole R, Dave N, Pai S, et al. As (III) removal using superparamagnetic magnetite nanoparticles synthesized using Ulva prolifera–optimization, isotherm, kinetic and equilibrium studies. Chemosphere. 2022;**308**:136271. DOI: 10.1016/j.chemosphere.2022.136271

[23] Zubair M, Roopesh MS, Ullah A. Nano-modified feather keratin derived green and sustainable biosorbents for the remediation of heavy metals from synthetic wastewater. Chemosphere. 2022;**308**:136339. DOI: 10.1016/j.chemosphere.2022.136339

[24] Mohamed MS, Hozayen WG, Alharbi R, Ibraheem IBM. Adsorptive recovery of arsenic (III) ions from aqueous solutions using dried Chlamydomonas sp. Heliyon. 2022;**8**:e12398. DOI: 10.1016/j.heliyon.2022.e12398

[25] Chen J, Sun J, Luo M, Li Y, Wang Z, Wang Y. As(III) oxidation and kinetic analysis by Herminiimonas arsenicoxydans-loaded electrospinning activated carbon fiber biofilms. Chemosphere. 2022;**30**:136479. DOI: 10.1016/j.chemosphere.2022.136479

[26] Pratama DE, Yang C-L, Chen YY, Lin C-C, Hu Y, Lee T. Quick-fix agarose beads impregnated with hydrous ferric oxide for As(III) species removal from pharmaceutical wastewater. Separation Science and Technology. 2022;**57**:1013-1029. DOI: 10.1080/01496395.2021.1962350

[27] Liu H, Li P, Qiu F, Zhang T, Xu J. The application of eco-friendly Fe–Al bimetallic oxide/biochar adsorbent composites with waste rice husk for removal of arsenic at low concentration. Journal of Inorganic and Organometallic Polymers and Materials. 2022;**32**:122-133. DOI: 10.1007/s10904-021-02085-z

[28] Behera US, Mishra PC, Radhika GB. Optimization of multiple parameters for adsorption of arsenic (III) from aqueous solution using Psidium guajava leaf powder. Water Science and Technology. 2022;**85**:515-534. DOI: 10.2166/wst.2021.613

[29] Wei Y, Zou R, Xia Y, Wang Z, Yang W, Luo J, et al. Enhanced arsenite adsorption from water by activated MOF embedded in macroporous chitosan bionanocomposite beads. Materials Today Chemistry. 2022;**26**:101091. DOI: 10.1016/j.mtchem.2022.101091

[30] Nakakubo K, Endo M, Sakai Y, Biswas FB, Wong KH, Mashio AS, et al. Cross-linked dithiocarbamate-modified cellulose with enhanced thermal stability and dispersibility as a sorbent for arsenite removal. Chemosphere. 2022;**307**:135671. DOI: 10.1016/j.chemosphere.2022.135671

[31] Ahmad I, Ghaffar A, Zakir A, ZUH K, Saeed MF, Rasool A, et al. Activated biochar is an effective technique for arsenic removal from contaminated drinking water in Pakistan. Sustainability. 2022;**14**:14523. DOI: 10.3390/su142114523

[32] Alguacil FJ, Escudero E. Removal of arsenic(V) from aqueous wastes by ion exchange with lewatit MP64 resin. Desalination and Water Treatment. 2018;**133**:257-261. DOI: 10.5004/dwt.2018.22676

[33] Agunwamba JC, Amu AM, Nwonu DC. An efficient biosorbent for the removal of arsenic from a typical urban-generated wastewater. Environmental Monitoring and Assessment. 2022;**194**:911. DOI: 10.1007/s10661-022-10631-6

[34] Akram A, Muzammal S, Shakoor MB, Ahmad SR, Jilani A, Iqbal J, et al. Synthesis and application of egg shell biochar for As(V) removal from aqueous solutions. Catalysts. 2022;**12**:431. DOI: 10.3390/catal12040431

[35] Maity S, Bajirao Patil P, SenSharma S, Sarkar A. Bioremediation of heavy metals from the aqueous environment using Artocarpus heterophyllus (jackfruit) seed as a novel biosorbent. Chemosphere. 2022;**307**:136115. DOI: 10.1016/j.chemosphere.2022.136115

[36] Ma L, Xu H, Xie Q, Chen N, Yu Q, Li C. Mechanism of As(V) adsorption from aqueous solution by chitosan-modified diatomite adsorbent. Journal of Dispersion Science and Technology. 2022;**43**:1512-1524. DOI: 10.1080/01932691.2021.1876592

[37] García-García JJ, Gómez-Espinosa RM, Rangel RN, Romero RR, Morales GR. New material for arsenic(V) removal based on chitosan

supported onto modified polypropylene membrane. Environmental Science and Pollution Research. 2022;**29**:1909-1916. DOI: 10.1007/s11356-021-15725-1

[38] Lee S, Han J, Ro H-M. Mechanistic insights into Cd(II) and As(V) sorption on Miscanthus biochar at different pH values and pyrolysis temperatures. Chemosphere. 2022;**287**:132179. DOI: 10.1016/j.chemosphere.2021.132179

[39] Din SU, Hussain B, Haq S, Imran M, Ahmad P, Khandaker MU, et al. Efficient arsenate decontamination from water using MgO-Itsit biochar composite: An equilibrium, kinetics and thermodynamic study. Watermark. 2022;**14**:3559. DOI: 10.3390/w14213559

[40] Jiang C, Zhang T, Li S, Yang Z. A comparative study on Fe(III)-chitosan and Fe(III)-chitosan-CTAB composites for As(V) removal from water: Preparation, characterization and reaction mechanism. Environmental Science and Pollution Research. 2022;**29**:77851-77863. DOI: 10.1007/s11356-022-20701-4

[41] Verma L, Azad A, Singh J. Performance of a novel iron infused biochar developed from Raphanus sativus and Artocarpus heterophyllus refuse for trivalent and pentavalent arsenic adsorption from an aqueous solution: Mechanism, isotherm and kinetics study. International Journal of Phytoremediation. 2022;**24**:919-932. DOI: 10.1080/15226514.2021.1985078

[42] Abboud AS, Mollahosseini A, Ghaffarinejad A. Alginate caged graphene oxide -modified metformin beads for the removal of arsenic(III) and (V) from aqueous media; kinetic and equilibrium, thermodynamic studies. Separation Science and Technology. 2022;**57**:2894-29072022. DOI: 10.1080/01496395.2022.2094275

[43] Chang Chien S-W, Weng C-M, Chou J-S, Liu C-C. Application of δ-MnO_2 and biochar materials in an arsenic-contaminated groundwater. Water Environment Research. 2022;**94**:10811. DOI: 10.1002/wer.10811

[44] Zhang J, Huang L, Ye Z, Zhao Q, Li Y, Wu Y, et al. Removal of arsenite and arsenate from contaminated water using Fe-ZrO-modified biochar. Journal of Environmental Chemical Engineering. 2022;**10**:108765. DOI: 10.1016/j.jece.2022.108765

[45] Suresh S, Sillanpää M, Banat F, Vissa RK. Adsorption of arsenic in aqueous solution onto iron impregnated bagasse fly ash. Journal of Environmental Health Science and Engineering. 2022;**20**:861-879. DOI: 10.1007/s40201-022-00827-w

[46] Ali H, Ahmed S, Hsini A, Kizito S, Naciri Y, Djellabi R, et al. Efficiency of a novel nitrogen-doped Fe_3O_4 impregnated biochar (N/Fe_3O_4@BC) for arsenic(III and V) removal from aqueous solution: Insight into mechanistic understanding and reusability potential. Arabian Journal of Chemistry. 2022;**15**:104209. DOI: 10.1016/j.arabjc.2022.104209

[47] Ahiahonu EK, Anku WW, Roopnarain A, Green E, Serepa-DlaminiMH,GovenderPP.Exploring indigenous freshwater chlorophytes in integrated biophotovoltaic system for simultaneous wastewater treatment, heavy metal biosorption, CO_2 biofixation and biodiesel generation. Bioelectrochemistry. 2022;**147**:108208. DOI: 10.1016/j.bioelechem.2022.108208

[48] Fauzi WN, Fairuz FSM, Shukri NM, NFM S, WNW A, NAM S, et al. Abdullah NH. Removal of As and Cd ions from aqueous solution using biosorption technique. IOP Conference Series: Earth and Environmental

Science. 2022;**1102**:012088. DOI: 10.1088/1755-1315/1102/1/012088

[49] Figueira P, Henriques BF, Afonso N, Pinto J, Tavares D, Vale C. Pereira Eduard. Potentialities of agro-based wastes to remove Cd, Hg, Pb, and As from contaminated waters. Water, Air, and Soil Pollution. 2022;**233**:78. DOI: 10.1007/s11270-022-05543-7

[50] Iwuozor KO, Akpomie KG, Conradie J, Adegoke KA, Oyedotun KO, Ighalo JO, et al. Aqueous phase adsorption of aromatic organoarsenic compounds: A review. Journal of Water Process Engineering. 2022;**49**:103059. DOI: 10.1016/j.jwpe.2022.103059

[51] Ryu J, Han Y-S, Cho D-W, Kim S-J, Cho Y-C, Chon C-M, et al. Practical application of PAC sludge-valorized biochars to the mitigation of methyl arsenic in wetlands. Chemical Engineering Journal. 2022;**450**:138148. DOI: 10.1016/j.cej.2022.138148